# Water & Wetland Plants

## FOR SOUTHERN AUSTRALIA

# Water & Wetland Plants
## FOR SOUTHERN AUSTRALIA

·Nick Romanowski·

LOTHIAN BOOKS

*This book is for Jan.*
*No matter how complicated our lives, she has always made sure the foundations are strong.*

A Lothian Book

Lothian Publishing Company Pty Ltd
A division of Thomas C. Lothian Pty Ltd
11 Munro Street, Port Melbourne, Victoria 3207


First published 1992



National Library of Australia
Cataloguing-in-publication data:

Romanowski, Nick, 1954–
Water and wetland plants for southern Australia.

ISBN 0 85091 487 6.

1. Water gardens — Australia. 2. Aquatic plants — Australia.
3. Gardening — Australia. I. Title.

635.96740994

Cover design by Geoff Hocking
Text designed by Geoff Hocking
Typeset in Century Schoolbook by DigiType
Printed in Australia by Macarthur Press

# *Acknowledgements*

A book of this kind could not have been written without the help of hundreds of individuals who have generously given information over the past five or so years. There are some who have contributed so much that they deserve special thanks; in particular, David Albrecht, Surrey Jacobs, Edith Metelik, Walter Pagels, Peter Rogers and Karen Wilson.

I would also like to acknowledge an early pioneer of watergardening, Amos Perry, whose encyclopaedic *Water, Bog and Moisture-loving Plants* was the first really good guide to wetland plants.

# Contents

# Introduction

Water is tranquil and mysterious. It can make any garden look out of the ordinary and provide a focus which ties a garden together, giving it a living heart. While even a bare formal pool has an aura which nothing else can provide, the atmosphere and feel of most watergardens (including the famous ones) is created by plants. Waterplants and moisture-loving plants define the texture and colour of the water; their growth and changes reveal the mood of each season. Yet in most watergardening books, plants are often relegated to an appendix at the back. Only a fairly small range of plants is usually covered, giving the impression that the few common plants generally available are all that there are. It isn't surprising that most watergardens look much the same; their owners have chosen from a selection of perhaps only 30 or 40 varieties.

This book gives information on around 600 species and varieties of waterplants and moisture-loving plants, all of them available to gardeners in southern Australia, and gives cultivation information for Australian conditions. In most cases, the plants described will grow from Tasmania to southern Queensland, and in the south-western corner of Western Australia; exceptions are described under the individual plant descriptions. I have avoided covering plants which will only grow in the more tropical parts of Australia for two reasons. Firstly, adding these would double the size of the book. Secondly, the native species of northern Australia are poorly known at present, and almost nothing is known about the cultivation of the new species still being described and studied.

I have also avoided describing *every* plant which likes constant moisture; many of these are woodland plants from climates with a much more regular rainfall than ours, and they will not tolerate soils which are really wet at any time. Trees and larger shrubs are also left out — there is already a considerable specialist literature on these, and they are not ideal plants for ponds or dams. Many trees, such as willows, will suck tremendous amounts of water out of a pool if their roots have access to it (of course, if the pool is completely waterproof it doesn't matter whether nearby trees are water-loving or not). The plants described in this book are almost all associated with water in nature, growing in rivers or beside them, in marshes and lagoons. Some are inhabitants of temporary pools or swamps and will also survive drying out at some times of the year.

A feature of this book is the coverage of Australian species, including new information on their cultivation. Although the cultivation of native waterplants has been neglected until recently, there is now a growing interest — this will not come as a surprise when you see some of the plants now available. Another unusual feature is the inclusion of well-known garden plants which are naturally associated with water but are also fairly tolerant of drier conditions; many such plants will only show best size, growth and flowering as waterplants.

Finally, this book sets out to make watergardening a part of contemporary gardening instead of the mish-mash of antiquated ideas it has become. The growing of plants in water and wet soils has not accumulated much more than a hundred years of history in western cultures; in that time, many myths have been spawned and copied from one watergardening book to another without any critical thought. Among the myths which are here discussed is that of the 'essential oxygenating plants', and the mysterious hybridisation programme through which the Frenchman Marliac created most of the hardy, hybrid waterlilies still offered for sale today.

If you're planning a watergarden, or want to make something more of one already established, this book offers you the key to cultivation of an unimaginable range of plants. There are countless millions of possible combinations of the varieties listed; no two Australian watergardens need ever look the same again!

# Chapter One

# Making Ponds and Dams

Pond design, construction and maintenance are covered in every book on watergardening ever printed, and some have covered these subjects very well indeed. This chapter is only an outline of the most important basics, but emphasises information which is not usually available elsewhere, such as dam construction, water quality, choice of fish for southern Australia, and a variety of potential problems.

## Planning a pond

The type of pond you choose to build will decide the materials you use, where it is located, and may limit the types of plant you will choose to grow. One important need for ponds is sunlight — most waterplants need at least four hours of direct sun, but six to eight hours is better still. Morning to mid-afternoon sun is best, especially over the hotter months, but any sunlight is better than none. Northern hemisphere gardening books recommend full sun, but this is not always desirable in Australia, especially if you wish to grow plants from cooler climates. As is the case with cottage-garden plants, many water plants cannot tolerate the full blast of a February sun in Australia, so try to arrange some shading either in the afternoon or around the middle of the day.

Small ponds can be made to suit even tiny areas such as verandahs; the simplest is a wooden half-barrel which must not have been used for anything more potent than alcohol or vinegar. (Those which have been treated with any kind of wood preservative must not be used.) Scrub the inside of the half-barrel thoroughly *without* soap before you fill it, then change the water every few days until little or no smell remains. If the pond is to go on a verandah or wooden floor, make sure this is strong enough to take the weight; a pond of this kind, measuring one metre in diameter, holds about 400 kg of water. The half-barrel must be kept full of water or it will dry out and start to seep, and the hoops holding it together should be secured with brass screws or they may slip.

Larger ponds can be formal, semi-formal or natural in their design, and your choice must fit in with the look of the rest of your garden. Neatly laid-out, regular garden beds would suggest a formal or semi-formal pond, while a wild garden would obviously have a natural pond. Whatever your choice, mark the outline of the planned pond on-site with a rope, and spend some time looking at the marked area from different parts of your garden. Make the pond as large as you can afford and have space for; few pond-owners haven't wished their pond was just a little larger once it was built. Ponds are not very moveable, so take your time choosing the location, and keep in mind that there are factors apart from appearance and sunlight which limit the selection of sites. For example, underground pipes and cables must be avoided, or you may need to have space for machinery to work in the area. And don't forget that a pond can be just as much of a trap for small children as a swimming pool.

Formal ponds should be kept as simple as possible, especially if they are small. They are placed as a focal feature for a formal garden, and plantings are kept simple to allow for maximum water surface and reflection. The best formal shapes for most gardens are rectangles, ovals, circles and semi-circles. Combinations of these geometric shapes, such as two circles linked by a rectangle, can be used for very large ponds, but may look cluttered and clumsy in smaller versions.

A formal pond

Semi-formal ponds are more loosely designed, and can blend a variety of formal shapes. They should also be of a reasonable size, and not include too many different forms in their outline. Natural ponds are the hardest of all to create; a really natural-looking pond is rare for a number of reasons, the most important of which is that not many people have seen truly wild wetlands on which to model their ideas. Unless your garden is a natural-looking bushland, you will not achieve a natural-looking wild pond — and may not want to. Some of the most beautiful ponds I have seen are semi-formal, blending almost natural-looking plantings together with simple, elegant shapes to give the feel of a well-planned garden without losing the appearance of an open expanse of water.

# Pond construction and materials

There are three basic materials used in pond construction: concrete, fibreglass and pond-liners. You will soon have the idea that I favour pond-liners over other materials, so I should say that I don't sell them (or any other kind of pond) myself, or have shares in any company which does!

I am sometimes asked how to make ponds in the English fashion, by 'puddling' clay. I don't recommend this method which was developed for a climate with reliable summer rains; it relies on good-quality water-holding clays which are worked by hand to give proper compaction and are not allowed to dry out once filled. Even a week of a gentle Australian summer sun can lower the water level of a small pond by 10 cm, and the water-holding ability of exposed clay will quickly be destroyed by drying out if you happen to be away for a week or two at the time.

A problem which must first be considered for any pond which is dug into the ground is what to do with the soil you remove. Even a small pond will result in an impressive mound of soil, which can either be removed (a major job in itself) or used to form a rockery or raised bed elsewhere in the garden. If you are planning to put in waterfalls or watercourses then this soil will obviously form the foundation for these, but it should otherwise be removed to its new site as fast as it is dug out.

## Concrete

Concrete ponds were once the only form of permanent pond, but the amount of work involved in preparing them, and their relative unreliability, has made them a rarity since pond liners became available. Concrete is still worth using as a pond material *providing* you are already familiar with its use and can do most of the work yourself. If you intend to pay someone else to do the work, the cost will be comparable with a pond made of the most expensive liners.

Many home-handyman books give good instructions for building ponds if you feel you are experienced enough to use concrete, but there are several points which must be emphasised:

■ All the concrete must be poured within a few hours or you will find that water runs out through the join between the old concrete and the new.

■ The walls and base must be reinforced with heavy chicken- or rabbit-mesh, and in larger and deeper pools the mesh should be further supported by concrete reinforcing mesh.

■ A 15 cm layer of compacted sand should be laid underneath the base of the pond in order to spread the pond's weight evenly.

■ If you are making thick walls using forms, a concrete vibrator is essential to eliminate air pockets (a guaranteed source of future problems).

■ Before use, the pond should be scrubbed with vinegar, filled with water, then allowed to stand for several days in order to neutralize free lime which can kill fish and plants (this process may need to be repeated several times).

Even if you follow all proper procedures, concrete ponds still have a lovable habit of developing invisible leaks or cracks which are difficult to find and seal properly. You may be able to repair such leaks for a year or two, but the same problem often recurs with time. If the leak is a fine hairline one, painting the concrete with a non-toxic swimming-pool paint may solve the problem, and you won't need to go through the long process of curing the concrete either.

To avoid further discouraging you from trying to make your own concrete pond, I won't even mention the lung and skin problems that cement dust can cause, the plumbing which may need to be built into the base, or the fact that concrete ponds which won't hold water are usually repaired with the pool liner you could have bought in the first place.

## Ferrocement

Ferrocement is another form of concrete which is particularly good for freeform ponds. Tie chicken mesh together in several sheets to make a support which can be pushed into the form desired, then work a thick slurry of ferrocement through the mesh with a trowel. The mix is usually half *fresh* cement, and half coarse sand. Like any other form of concrete this must be cured before the pond can be used.

Although ponds made this way are very strong and long-lasting, a ferrocement pond may cost as much as a comparable pond made with an expensive liner — even if you do the work yourself — due to the high proportion of cement used.

## Fibreglass

Fibreglass ponds are readily available from general nurseries in a range of sizes, shapes and designs, most of which are unattractive, inconvenient to install, and no cheaper than good-quality pond liners of a comparable size. The main reason they are sold appears to be that some nurseries know very little about the newer liner materials. The hole for a fibreglass pond must be larger than the pond itself, which is bedded on a layer of packed sand, levelled precisely, then backfilled to hold it in place.

I'm not completely against fibreglass ponds; in fact, nearly all the ponds in my nursery are fibreglass. However, these are heavier-duty, free-standing rectangular ponds which are not designed for beauty but for the most efficient growing of waterplants. An advantage of fibreglass is that it can be quickly and easily repaired with silicone sealant, even when quite severely damaged.

## Pond-liners

For ornamental ponds which are dug into the ground, I would never use anything except pond-liners. There are a number of long-lasting materials now available for holding water. My favourite has been butylene rubber, a synthetic with a probable lifespan of 50 to 80 years, which was originally developed to hold potent chemicals during the Second World War. Other materials are also becoming available, many of which are cheaper than butylene yet are still offered with very long warranties. Some of the dearer products are very long-lasting and have guarantees of 20 or so years — well worth paying extra for. When you are deciding which material to use, obtain information on each product from the retailer and compare their properties. Properties you should compare include colour, stretch (it is easier to lay a very elastic liner than a stiff one), and stability to ultraviolet light. Black is the only shade which gives a reasonably natural- and deep-looking pond, but you may have some other effect in mind with blue, green or other coloured liners. Avoid brown liners; all the ones I have seen give a dingy effect quite unlike the colour of soil, and seem to make fish and plants look less colourful.

There is no need to describe the many ways liners can be used to build ponds, as competition between manufacturers has resulted in scores of booklets and pamphlets describing every possible method. Obtain and read as many of these as possible before you even decide what sort of pond you would like to build. A pond between 3000 and 5000 litres in volume can be dug by hand in one or two days. Larger ones can be made even faster if you use machinery to do the digging.

## Waterfalls

Waterfalls and watercourses are also commonly included in ponds; these can also be made with sheets of pond-liner hidden below rocks and gravel. It is easy to make an unnatural-looking waterfall; the most common fault I have seen is one whereby the watercourse is obviously higher than its surrounds. This is best hidden with vegetation growing on either side above the level of the moving water, or by having a relatively simple facing of rock down which water flows. For a formal or semi-formal watercourse, there is no need to attempt to hide its artificial nature. Some of the finest examples I have seen of this kind simply race water through rectangular channels, dropping over unornamented and undisguised walls from one level to another. For maximum waterfall commotion, the water can be run over an overhang before falling to a pool below, and even a very small water flow can give dramatic results in this way.

Underwater lights, pumps, filters and other accessories are beyond the scope of this book, for the simple reason that so many new brands and variations of these have appeared recently in Australia that a section describing them would be out of date in a few months. Most garden-related journals carry articles describing new developments of this kind, and the only way to know what is currently available is to read them regularly.

However, there are two general warnings which should be given about some of the electrical novelties offered. Firstly, not all of these devices are necessarily approved for use in Australia as they may not meet safety standards. Electricity and water aren't good mixers, so make sure anything you buy has been approved in your state, and also for the use you want to put it to. Secondly, some of the newer brands now proliferating are not going to stand the test of time. As was the case with a similar glut of aquarium products a few years ago, some will prove to be short-lived, and spare parts and repair facilities may not be available when you need them. Check guarantees and availability of parts; is there an Australian stockist who *already* keeps a full range of spare parts, and is qualified to service this brand? Most manufacturers of good quality products provide information booklets on their products which will answer these questions. Read them carefully before you buy.

## Planning a dam

The first priority in planning a dam is deciding what you want to use it for. Stock-watering and keeping large numbers of wildfowl or domestic waterfowl are not compatible with ornamental plantings or swimming, nor are they ideal for aquaculture. It is assumed that most people reading this book will have a particular interest in planting, which can be combined with swimming, fishing and some kinds of aquaculture. All these activities require water which is preferably fairly soft and clear, and not too enriched with nutrients from run-off or animal manures.

To maintain reasonable water-quality, the catchment area directly above the dam should be planted quite densely to prevent soil and excess nutrients — two factors which contribute to the muddy or discoloured appearance of many dams — from washing into the water unnecessarily. Although many water and wetland plants will help to keep the water clear, the first filtering stage is always the land plantings — these reduce the amount of silt and nutrient which can get into the dam in the first place. If you don't have time to allow plants to grow and bind the catchment together, it is a good idea to install a silt trap or settling dam; this slows down the water entering the dam, giving it time to drop most of the mud and detritus which heavy rains may otherwise deposit in the main dam.

As a dam or lake relies on natural run-off to keep it filled it must be located where there is enough run-off to fill it. The water level is likely to drop (through evaporation) over the warmer months. Even in the wetter parts of southern Australia evaporation losses can be as high as 1 m over a dry summer. Not all waterplants will tolerate such changes in water level, so you will either need to plant species able to tolerate such variations, or have another dam (preferably sited at a higher level so water can be moved by gravity) which can be used to top up the main dam. If the natural catchment of the dam does not fill it with water in an average rainy season, you may need to have channels running outwards and slightly uphill from the dam in order to redirect additional run-off which would otherwise escape.

It is important to work out the exact level a future dam will reach; this is done with a surveying level. Mark the planned height of the dam *at the overflow* (not at the top of the wall which is always higher), then use the level to peg out the outline of the area the dam will cover. Alternatively, you can mark out the outline of the area you want filled, and work out the height of the dam wall and the overflow from this. At this stage, it should become obvious where islands, peninsulas, planting shelves and shallows can be made with minimal earthmoving. Apart from being more economical, minimal earthmoving also gives a more natural appearance. Some people attempt to economise further by planning to build floating islands, but I have yet to see one which looks attractive or natural no matter how much work has been put into it — such islands are usually built as refuges for waterfowl rather than for ornamental plantings. Don't overdo the number of islands and peninsulas; few dams look remotely natural with anything more than a single island, and one or two peninsulas.

# Dam construction

Dam construction follows the stages described below, some of which appear to be unknown to many of the people who build them. One accessory which *must* be used is a sheepsfoot (also called a calfsfoot) roller. Operators who construct dams with only a bulldozer will assure you 'she'll be right', and indeed, 'she' may well be two times out of three, but the sad fact is that one in three dams in Australia doesn't hold water properly because of poor construction. The use of a sheepsfoot roller is essential; bulldozers run on caterpillar treads designed to spread the weight of the bulldozer. These treads do such a good job of distributing weight that people have been run over by bulldozers without serious injuries — a sure sign of how poorly they compact soil.

All the steps described below should be followed unless you have *expert* opinion that some can be left out because of your local conditions.

**Step 1:** First, remove all topsoil from the areas to be excavated and stockpile it as close to the site as possible without it getting in the way of work.

**Step 2:** The base of the future wall should then be ripped (loosened) to a depth of about 5 cm to allow new material added on top to tie-in with the foundation. Deeper ripping will leave a loose layer of subsoil which won't compact properly, and will allow water to seep through later. An alternative to the ripping-and-bonding method is to dig a core trench into the subsoil below the future wall, then fill it with good-quality compacted clay; continue building and compacting on this foundation. This method is only used on very large or deep dams where it can save construction time and reduce overall cost.

**Step 3:** If you need to put a pipe through the dam wall, do so before construction of the wall begins. A trench must be dug below the ripped area, and the pipe laid in this, then the area around it compacted thoroughly by hand before the wall is built above it. Water will seep along any weakness running through the dam wall whether it is a pipe, a tree root, or even just a piece of wire. To prevent seepage of this kind eroding around the pipe, baffle-plates must be firmly sealed around the pipe at regular intervals to stop water from flowing along the pipe to the point where it comes out of the wall. However, unless you need to be able to draw more than 70,000 to 100,000 litres of water out of the dam in 24 hours, a permanent siphon of about 3 cm diameter with a ball valve to close it off at the bottom is much cheaper than a baffled pipe. Such a siphon can be installed quickly by one person, and does not build a potential weakness into the wall.

**Step 4:** Subsoil is then excavated from the future dam area, and used to build up the wall. If much of the subsoil is unsuitable for a water-retaining wall, you may need to line the entire inside of the dam with at least a 60 cm-deep layer of better-quality clay to seal it. However, soils with as little as 20 per cent clay content are suitable for a water-retaining dam wall if properly rolled by a sheepsfoot roller.

**Step 5:** Towards the end of the wall-construction phase, you can have islands, peninsulas, planting shelves and shallow areas cut in. You may also have to start rolling the wall with the sheepsfoot roller before the wall is fully built up, depending on how deep it is; for every 3 m of depth, the wall should be rolled until the roller 'walks out', i.e. the spikes no longer sink deeply into the wall material, but run across its surface. For every additional 3 m of depth, another 15 cm-layer of material should be added and rolled in the same way.

**Step 6:** The dam must include an overflow below the level of the top of the wall which is wide enough to allow the heaviest flooding likely to occur in your area to escape without running over the top of the wall itself. The overflow is usually located at the side of the dam. Water running through the overflow can be a major erosion problem which may weaken and undercut the wall, so the run-off area should be paved with coarse rock, terraced, concreted or otherwise prepared to withstand the effect of flood conditions.

**Step 7:** Finally, topsoil is replaced on islands and peninsulas and over the planting shelves and shallows which will later be under water (these are discussed in more detail in Chapter 2). If you need to mix anything into the topsoil (e.g. manures and other fertilisers), a skilled bulldozer operator can do this job for you before the soil is spread. If you enrich the soil you have spread on the planting shelves or if it is naturally very rich, a 5 cm-deep layer of river gravel or similar material should be spread

Jetty

over the soil to reduce nutrient loss into the water once the dam fills. Soil should not be spread over the deeper sections of the dam unless you have a specific reason to have it there, as silt will tend to accumulate there with time anyway, and there is no reason to speed up the process. Information on planting shelves and soils for underwater use is included in the next chapter.

If you plan to spend much time near your dam in summer, I cannot recommend too highly the construction of a jetty. Put the foundations in before the dam is filled (I have done this job underwater and don't recommend it); the rest of the framing can be finished at any time above water. A slatted canopy over the top will keep the worst of the summer sun off your skin while allowing creepers to grow through the roof as permanent shade. Jetties are the perfect base for swimming, fishing, or just relaxing. Make sure you include solid hooks in the frame to hang a hammock in the shade.

# *Water chemistry*

Pure water is the most harmless of all liquids, yet it is rarely found in nature. Most natural water contains other substances which change it and affect the plants and animals living in that water. Many solids will dissolve in water, and there are usually several gases also present. The role of gases is fairly straightforward, so let's look at these first.

## Dissolved gases

There are only two gases normally present in water which are particularly important to living creatures: oxygen and carbon dioxide. Oxygen is the gas animals must breathe to 'burn' food chemicals and release energy, but plants also use it at night. Carbon dioxide is the gas produced once the oxygen has been burnt; it is a waste product from the point of view of animals, but plants use carbon dioxide during the day and break it up to release the oxygen again. The warmer water is the less gas it will hold, thus cooler waters hold more oxygen. If you stock a pond with as many fish as will survive in average conditions, they may rapidly die if the water becomes much warmer and the amount of oxygen it holds becomes less. If you want to stock a pond heavily with water-breathing animals, you must be able to maintain extra oxygen in warmer weather by circulating the water (but keep in mind that too much water movement will affect a number of waterplants, including waterlilies which will not do well in fast-flowing water).

It is often recommended that ponds be stocked with so-called 'oxygenating' plants which are supposed to add oxygen to the water. This is a myth left over from the dawn of watergardening and is partly based on confusion between the need for oxygen and the effects of carbon dioxide. Firstly, let's look at the effect of carbon dioxide. It is more than just a harmless waste-product in water. If there is too much of it, it is actually harmful. Water will hold much larger amounts of carbon dioxide than it will of oxygen. When fish are seen gasping at the water's surface, too much carbon dioxide is often the cause — because there may still be a sufficient amount of oxygen dissolved. The cure is exactly the same as for a lack of oxygen; either circulate the water to get rid of waste gases, or reduce the number of underwater animals if overcrowding is the problem. Although plants will remove carbon dioxide during daylight, at night they create more of it.

Using *Elodea* and *Egeria* as examples, let's consider the oxygen-production of plants. These two plants are supposed to be among the best oxygenating plants available (they are also potentially serious weeds). Part of the evidence for this is that both will produce small bubbles of oxygen when kept in a warm, sunny spot. However, warmer water holds less of any gas than cooler water, and the bubbles which are seen rising under such conditions are in fact oxygen which cannot dissolve. Although some of the oxygen will enter the water, this display of oxygen-production is largely wasted. In much cooler conditions (when the water will absorb most of the oxygen the plants produce) not much oxygen is made available at all.

In the narrow confines of aquaria, maintaining an adequate amount of oxygen has always been a priority, and it has been known for many decades that waterplants don't make any difference to the stocking capacity of an aquarium. On the other hand, a fast-moving stream of bubbles can increase the carrying capacity of the same aquarium by an estimated three times. It was originally believed that gas exchange between the bubbles and the water was the source of the additional oxygen, which implied that the bubbles produced by *Elodea* and *Egeria* could have a similar effect (especially as they are pure oxygen, not air). However, when research on this problem was finally done, it was found to be not the bubbles themselves, but the currents set up by the bubbles which speed up gas exchange; most of the gas exchange is through the moving surface of the water. The bubbles of oxygen produced by waterplants are too small to have a similar effect. In any dam or reasonably large pond the slightest breeze will also result in currents which turn over the water — this is enough for a reasonable exchange of oxygen and carbon dioxide through the water's surface.

In smaller or more sheltered ponds where air movement doesn't circulate the water, a dense growth of plants will actually reduce oxygen levels at some times. The reason for this is that in darkness plants *use* oxygen instead of producing it. Therefore, in a densely-planted small pond during warm weather, plants use up oxygen overnight and water-breathing animals can suffocate as a result, or at least suffer discomfort from the carbon dioxide which builds up as a by-product of the oxygen-use of the plants in darkness.

In the relatively cold conditions of a European summer, water won't often reach temperatures which cause severe depletion of oxygen, and plants may not greatly reduce oxygen levels except under unusual conditions. In Australia, I would suggest that not more than around one-third of the volume of a smaller or more sheltered pond should be taken up by underwater plants.

This problem rarely occurs in dams and larger waterbodies as much of the water's surface remains free of waterplants and air movement creates enough circulation. However, during long periods of very hot weather, a surface layer of very warm water may form which cuts off oxygen supply to the cooler water below in deeper ponds and dams. This can be overcome by mechanically circulating the water (referred to as aeration); the most basic method is to simply swim freestyle for one to two kilometres per million litres of water in a dam! As swimming is a particularly gentle exercise perfectly suited to the kind of weather which causes this problem, both you and your watergarden will benefit immensely. Other methods include churning the water with a stationary outboard motor, or agitating the surface with a high-pressure jet of water from a hose (a fire pump may be necessary for this purpose in a dam). Aeration is most useful towards evening as low oxygen levels are generally most severe overnight, but further aeration early in the morning may also be useful.

Despite my scepticism about plants as oxygenators, they do have a number of useful functions in ponds which are discussed in more detail elsewhere. For example, finely-divided plants are an important attachment site for the bacteria which break down poisonous, nitrogenous wastes — the more plants that are present, the more of these bacteria there will be. Plants also use some of the by-products of animal metabolism once these have been processed by bacteria and fungi. Their importance as security and shelter for underwater animals cannot be over-emphasised either. Stress is a major cause of early death for animals (whether directly or by reducing resistance to disease) just as it is for humans.

Plants and animals aren't the only users of oxygen in a pond. Many tiny lifeforms, including bacteria, are also present in a pond and use oxygen. Under normal conditions, they aren't a problem, but if there is too much organic material in the water from uneaten fishfood, dead animals, etc., their number will increase in such proportions that there may be insufficient oxygen left for anything larger. This is a problem which is easy to prevent — don't let anything decay in the water and it won't happen.

## Dissolved chemicals

A number of chemicals are found dissolved in water, many of which change the nature of the water in strange ways. Looking at these effects,

I will simplify most of them because we are only concerned with the effects on living things, and not in the details of the chemistry itself.

## Hardness

There are different kinds of water hardness, but from a watergardener's point of view hardness is just a measure of the total amount of particular types of chemicals dissolved in water. In fresh water, the most important hardness chemicals are based on calcium and magnesium, but sodium-containing chemicals (including common salt) may also be present. Water containing a fair amount of salt is called 'brackish', and only a relatively small number of waterplants can tolerate brackish conditions; these same plants are often very tolerant of hard water, too.

Hardness can be measured with fairly simple water-testing kits sold in any aquarium shop. Their accuracy depends on the quality of the kit, but even the cheapest ones give a reasonably good idea of hardness. If you are using mains water for filling ponds, you won't necessarily need to do the testing yourself; the water trust providing your water should be able to supply you with this information, including how much the total hardness varies with the time of year.

The chemicals that account for hardness don't vanish as the water evaporates — they simply become more concentrated, and so hardness increases. If your water supply is moderately hard to start with, your pondwater will become harder and harder as you top up any water that evaporates. The easiest cure for this is to drain off part of the water and fill the pond with fresh water two or three times over summer.

The scientific way to measure hardness is in 'parts per million' (ppm). There are other systems which are still sometimes used, such as German or English degrees of hardness. Converting between the different types of measurement is a task which can make the strongest weep, so I will use ppm only. A measurement in ppm tells you how many parts of dissolved chemicals there are to a million similar-sized parts of water. Most waterplants and animals are happiest in fairly soft water ranging from 0 to around 150 or 200 ppm. Water in the range of 200–400 ppm is reasonably hard, and above this range few living things will thrive.

## The pH scale

Some of the chemicals present in water make it more acidic or alkaline. Battery acid is an example of a strong acid; caustic soda is an example of an alkali. If you only added a few drops of one or the other to a large amount of water, the water would only become slightly acidic or alkaline, which wouldn't worry most living things. However, greater amounts would cause the water to become increasingly unpleasant for living things, and then dangerous.

There is a scale used to measure degrees of acidity and alkalinity: the pH scale (an abbreviation of a chemical term). The scale begins at seven, which is the centre and is neutral (neither acid nor alkaline). Distilled water is neutral, and additions of anything acid will make it acidic. The more acid is added, the further *down* the scale from 7 the pH measurement goes. Thus a pH reading of 4 is more acidic than one of 6. Adding anything alkaline will push the reading *up* the scale from 7. Hence a measurement of 10 is more alkaline than one of 8.

The scale is not linear like a ruler, but logarithmic, which means that each step down the scale is actually *ten times* the last step. In other words, while a pH reading of 6 is a bit on the acidic side, a reading of 5 is ten times more acidic, and a reading of 4 is one hundred times more acidic than a reading of 6. Going up the scale from 7 works in exactly the same way, with each number ten times more alkaline than the last.

From the watergardener's point of view it doesn't really matter why the pH scale is set out in this peculiar way; the explanation is only of use to chemists. However, you need to know the actual pH reading of your water because different plants and animals have a preferred pH range, and much beyond that can stress or even kill them. Most plants will live comfortably at a pH range of 6.0 to 7.2 (slightly acid to slightly alkaline) and can tolerate small variations beyond this range.

Taking pH measurements can be done accurately enough with a cheap pH test kit available from most aquarium shops. The test liquid (usually bromothymol blue, which measures from 6.0 to 7.6) is added to a water sample which will then change colour. By comparing the colour to the chart supplied with the kit you have the reading. If you want to measure a wider range, a liquid called Universal Indicator may be obtained from some aquarium shops or a chemist, but it is harder to obtain fine readings around the neutral mark with this. Before taking readings it is important to rinse your sample container with the water you are about to test, preferably several times.

Hardness and pH are related in some ways. The chemicals which cause water hardness are also alkaline, so hard water is usually alkaline as well. On the other hand, acidic waters are usually soft because most acidifying chemicals don't contribute to water hardness.

## The nitrogen cycle

Animals which live in water also excrete in water, and as they also breathe in water, their waste can be a problem. While this doesn't seem to be the most sensible arrangement, we need to understand what happens to the waste.

Some waste products are fairly harmless and are rapidly broken down by bacteria and fungi. The most dangerous ones are nitrogen-containing wastes which are poisonous to fishes, as well as some other aquatic animals. Under normal conditions, a cycle involving two types of bacteria develops to reduce the more poisonous substances into fairly harmless ones which plants can use up. As ammonia (a result of the breakdown of urine and urea) builds up from fish wastes, a type of bacteria equipped to use ammonia also starts to build up and use these wastes. This type of bacteria is found naturally in all water and doesn't have to be specially introduced. While using ammonia these bacteria in turn produce another type of poisonous nitrogenous compound, nitrite, which is dealt with in the same way by yet another type of bacteria. The end product of this cycle is nitrate which is fairly harmless to most kinds of fish, and can be used as a fertiliser by most waterplants.

The build-up of ammonia can reach quite a high level over several weeks before enough of the right bacteria appear to reduce it. The same applies to nitrites, as the bacteria which break these down don't start to build up until there are enough nitrites present from the previous stage of ammonia breakdown. All in all, the whole cycle can take weeks or months to reach a balance depending on temperature, and over that span of time a smaller pond can become moderately poisonous to fish.

The same levels of poisons are unlikely to build up in a large pond or dam unless it is grossly overstocked in the first place. However, a small pond can become quite unpleasant for the animal inhabitants unless you stock only a small number of fish until the bacterial cycle is established. Even if you plan to stock a pond heavily after this stage, you should add new fish gradually so as to give the bacterial cycles time to catch up with the extra wastes.

Although nitrates are reasonably harmless to most fish, they can build up to an unacceptable level and cause toxicity. The cure is to either change part of the water regularly, or include a reasonable number of plants (which use nitrates directly). Ammonia, nitrite and nitrate test kits are available, but most are expensive and fiddly to use. You shouldn't really need them as long as you stock smaller ponds gradually.

## Turbidity (murky water)

Water may be murky or cloudy from a number of causes. Silt or other tiny particles may be suspended in the water, and if fish, yabbies or livestock keep stirring them up they may never settle. Water can pick up colouring from some clays or tree bark and leaves, and have a distinct colour which is almost impossible to get rid of. Fortunately, it is often an attractive and interesting colour, and one which leaves the water clear enough to see into.

Green, brownish or grey water is usually the result of an excess of nutrients, including fertilisers and waste products from animals; these colours can be a sign of ill-health or potential problems and are discussed under *Problem solving* on page 33. In the drought which ended in late 1991 blue-green bacteria (once called algae) built up to dangerous levels in some catchments because of high levels of nutrients; these cause further problems, as they produce poisonous waste-products.

## Chemicals deliberately added to water

Tapwater in Australia has usually been treated with a variety of chemicals. Fluoride is often added in the form of a chemical compound which was once used to poison rats before it was decided that it is good for the dental health of children. While it is no better for the general health of fish than it is for you, it is not necessarily dangerous in the small amounts added.

Chlorine is also added to tapwater as an antibacterial agent. It simply evaporates out of the water if the water is left to stand for one or two days, so don't add too much new water directly to an established pond as it can stress fish. Another chlorine-based substance called chloramine is in use overseas. If this comes to be included in the tapwater where you live, you will have to use chemical treatments to eliminate it because chloramines are very poisonous to fish and do not evaporate. Most aquarium shops sell water treatments which will eliminate chloramine, along with chlorine, but make it clear that you plan to treat a pond with it as packs designed for use in aquaria are too expensive for use in large volumes of water.

# *Fish for ponds*

Fish have a number of uses in ponds; they are both ornamental and entertaining, and they stop pests such as mosquitoes from breeding. While many small fish are suitable for ponds, one you should never add to a pond is the introduced mosquitofish (*Gambusia affinis*). They are the underwater equivalent of a cross between a rabbit and a dwarf piranha and although they were introduced to control mosquito wrigglers, they don't actually do a better job at this than any other fish of a comparable size. They will kill other smaller fish, damage the fins of larger ones, and breed like flies. They are attractive enough, but if you must keep them, do so in a small aquarium where you can see them and not in your pond where they will be difficult to eliminate. Most importantly (regardless of the types of fish you keep), never release them into natural waters. The exception here is native species which can be returned to the same waters you caught them from — but not elsewhere.

One of the most commonly kept fish is the goldfish, which comes in a number of deformed varieties such as Black Moors and Orandas as well as the normal-shaped varieties. Goldfish are messy feeders and stir up silt and detritus from the bottom of the pool, so if you want clear water without having to use filtration don't keep them. They will breed in ponds but many of the offspring revert to the colouration of wild fish, a bronze or silvery-green. As this colour is almost invisible in a pond, you will need to catch all the wild-type fish out each year, or you will end up with well-camouflaged fish after a few generations.

Comets and Shubunkins are the hardiest goldfish, and are always basically fish-shape, with just a single tail of variable length. Comets vary in colour from orange, gold and white through to deep reds in some choice strains. Shubunkins have transparent scales through which various underlying colours can be seen — the best strains are multicoloured in blue, white, orange and red, splashed with black spots and shining, metallic scales.

The double-tailed goldfish varieties have rounder bodies than Comets and Shubunkins, so they are poor swimmers and many develop digestive or swim-bladder problems in cold weather. This is the result of stress on a body which is deformed to a sufficient degree to cause ill health. The combination of intolerance plus the ease with which these clumsy swimmers can be caught makes varieties such as Fantails, Orandas and Black Moors unsuitable as pond fish except in relatively warm climates where the fish can be protected from cats, birds and small children (their major predator).

Koi carp are legal to keep in some states, but not in others. They grow far too large for most ponds and prefer deeper water than most plants will tolerate. In any case, they will uproot and damage plants, and should be kept in large, unplanted pools with special filter systems to clean up the remarkable amount of waste they produce while feeding.

Other smaller, non-native fishes which will live in ponds include Bitterlings and Weather loach, but these disappear from sight the instant you put them in, never to be seen again. Zebra danios are not hardy in the coldest climates; however, they will school and be fairly conspicuous if your winters are not too severe. White Cloud Mountain-minnows are one of my favourite pond fish; males will spar in sunlight over summer, showing their red fins and glittering, neon-green stripe to advantage. Young fish are amazingly brilliant; the stripe along their side is a metallic blue-green which puts tropical fishes such as neon tetras to shame! However, White-Clouds will not thrive where dragonfly larvae (mudeyes) are common, as they haven't a trace of the common-sense most other fish show in avoiding these voracious little monsters.

Only a few native fish species are easily seen in ponds, and then only when it suits them to be seen. Several species and forms of rainbowfish (*Melanotaenia*) available in aquarium shops will survive outdoors in much of southern Australia if acclimatised gradually. The further north you live, the more species are likely to succeed. In the south, *M. duboulayi* (from eastern coastal rivers) and *M. fluviatilis* (from the Murray river system as far south as Seymour in Victoria) are the best prospects for self-reproducing populations in your pond. Rainbowfish are at their best in sunny ponds with dense, floating vegetation where they can hide, and open water near the plants where they will display their dazzling array of colours from reds and yellows, through to an assortment of metallic blues and greens.

Dwarf galaxias (*Galaxiella pusilla*), Pygmy perches (*Edelia* and *Nannoperca*), and various miniature gudgeons (*Hypseleotis, Philypnodon,* etc.) will live happily and breed in ponds also, but they are mysterious dwellers of the depths and are only seen by the most patient observers. If you have that patience, these fish are rewarding to keep and may lead you further into the fascinating study of our smaller, poorly-known native fishes. A number of these smaller species have vanished from the wild, and new regulations on which ones can be kept in captivity have come into force in some states, and are being considered by others. Before collecting any such fish from the wild, check the current regulations with the appropriate fisheries authorities as these are likely to change considerably over the next few years.

## Fish health

Fish health is a complex subject, but can be simply summarised by pointing out that diseases usually only occur when the fish are under stress. Newly-caught fish are particularly prone to stress and should be added to a pond with care. Mix pond water gradually into the water they arrived in, then allow them to swim out into the pond of their own accord. If there is any possibility of disease being introduced to the pond, a period of quarantine and acclimatisation in a well-planted aquarium (lack of shelter causes unnecessary stress) or wading pool is a good idea, and any medication which is necessary should be administered there rather than in the pond itself.

Feeding pond fishes is simplicity itself. If you don't put too many in, they will find all the food they need in an *established* pond, including those mosquito wrigglers you don't want. Many native species will need live foods (waterfleas, worms, etc.) added if you keep enough fish to strain the resources of the pond, but goldfish will happily eat most artificial diets. If you are feeding your fish an artificial diet, be careful not to overfeed as uneaten food decays on the bottom of the pond and fouls the water. Remember also that fish become less active and eat less in cold weather. Rather than risk having food decay, it is better not to add any food at all for the coldest months — the health of your fish won't be affected in the slightest, unless they are overcrowded.

# *Fish for dams*

Many smaller native fishes are suitable for dams; on the other hand, none of the smaller non-natives listed under 'Fish for ponds' is. Goldfish will revert to the wild colouration, leaving you with a multitude of almost-invisible, brownish things swimming in a sea of mud which they themselves stir up — and there is no way to get rid of them. Koi carp are even worse offenders in this regard. The other non-native species already mentioned will either fail to survive (White Cloud Mountain-minnows, Medakas, Bitterlings) or do too well and escape into natural waters (Weather loach, Mosquito fish).

Of the larger non-native fishes, Redfin are also undesirable in dams. Apart from the likelihood that they are carriers of a serious, virus-like disease which can affect native fishes (and which has led to bans on their further distribution), they breed prolifically at a small size in dams, and these stunted fish are quite useless as food. As they have no other qualities to recommend them, leave them out. Trout, on the other hand, either rainbow or brown are satisfactory food fish and grow well in dams which are too deep and cold for native species — in fact, they should not be kept in dams that are less than 3 m or 4 m deep as the summer heat will kill them if they don't have deeper waters to retreat to.

Any smaller native fishes which will breed in a dam are suitable for stocking, and a variety of these will provide a source of food for larger, carnivorous species. If possible, choose species which feed and breed in different locations for maximum populations (e.g. some prefer to feed over deeper, open waters, while others prefer to live and feed amongst dense vegetation). Local species and varieties are usually the best choice because they are more readily available, can be transported quickly, and it will not matter if any escape into local rivers should the dam overflow in heavy rains.

Larger native fishes suited to most dams in southern Australia include Silver perch, Catfish (*Tandanus*) and, to a lesser extent, Golden perch. All are good-sized edible species, and catfish may even breed if the conditions are to their liking. In colder climates River Blackfish are a good choice if you can get them — although small, they are an excellent eating fish. Murray cod and close relatives are not a good choice for any dam as they don't seem to produce much in the way of eating-size fish no matter how many you stock. This may be because they are cannibalistic, so there is the potential for ending up with just one very large fish even if you started with hundreds. Although Golden perch can behave in a similar way, they are much less likely to as they feed on a wider variety of smaller animals than do Murray cod.

Two final points are worth noting if you plan to stock larger fishes in your dam. Firstly, I don't recommend supplementary feeding for fishes in dams unless you have a good knowledge of what is going on below — it is very easy to pollute the water if you don't know how much of the food is being eaten. Secondly, there are regulations and restrictions in most states as to which fishes you can keep (both native and non-native). These have been very changeable in recent times, so check with your state fisheries for any permits or limitations that would apply to stocking your dam.

# Waterbirds

Waterfowl (ducks, geese and swans) are not generally very compatible with waterplants. A small number of birds on a large area of water will not cause much damage if the plants are already well-established, but a large flock of most types of ducks or related birds will turn the finest ornamental lake into a mudhole. Domestic waterfowl are all fairly large and destructive. Most domestic ducks are descended from the wild mallard, and like that species will pull plants apart in their search for food amongst the leaves and roots. Muscovy ducks are more closely related to geese than ducks, both of which cause havoc in their search for food and also eat many plants. Paradoxically, geese, although vegetarian, will only eat some plants, leaving many others well alone. They may, however, acquire a sudden fancy for a plant they don't want to eat but will pull it apart apparently out of curiosity. (Such plants are almost always prized specimens which are about to flower for the first time.)

A number of non-native ornamental waterfowl are kept in Australia; I would recommend that you visit a breeder of such species to admire the clay floors and chocolate-coloured water in their enclosures before putting them into your watergarden. It is mainly due to overcrowding of waterbirds that few plants other than hardy trees will grow under these conditions; one exception is the well-planted enclosures used for mandarin ducks which are so rare, expensive and shy-breeding in this country that they are often given greater space and cover than more common species, allowing a reasonable plant-cover to survive. The same effect should be possible with any larger waterfowl; keep relatively small numbers of waterfowl on large areas of water with lots of well-established vegetation, and both plants and birds will thrive.

Native waterbirds may be divided roughly into four groups in terms of their compatibility with planted waters. The first group, including wood ducks (also called maned geese) and shelducks actually have little to do with water much of the time. These species nest high in trees, and usually feed over grass. They require access to pasture, and need water mainly for raising their young and as shelter during their migrations.

The second group feed almost entirely on submerged vegetation and grasses; the best example of this group is the black swan. Although they are destructive of plants, leaving wracks of vegetation to wash ashore and rot, they will not stay long on most dams simply because there is not enough food of the type they require. If you feed them with supplementary foods to encourage them to stay, they may breed on your dam if a suitable nest-site is available. In this case you may be able to enjoy being beaten with their powerful wings, and bitten should you try to swim or otherwise enjoy your dam in their breeding season. Fortunately, breeding is most likely to occur in the colder months.

The third group includes the common black duck, which is closely related to the introduced mallard (a number of so-called black ducks I have seen offered for sale are actually hybrids of the two species) and just as destructive to plants. While small numbers on large, well-established waters will cause little harm, they are no better on small ornamental dams than domestic ducks. Smaller related species, including chestnut teals, are far more desirable residents — although they are too destructive for smaller ponds, a pair or two may be quite compatible with a well-planted, large and established watergarden.

The fourth group that may visit your dam are predators on fish. How you regard them depends on how you value your fish. In a dam with well-established populations of smaller fishes, birds such as herons, egrets and cormorants will mainly feed on these smaller fish. Larger birds such as pelicans can catch much larger fish; however, they will be such rare visitors (except if your dam is close to the sea) that it is worth losing a fish or two to them for the pleasure of their company. Predators should be discouraged when you have recently introduced a large number of young fish to the dam. Chasing predators away for a few weeks at this stage will give the new arrivals time to settle in and learn to evade the regular dangers. Some people shoot cormorants, the most serious and persistent of the flying predators, but as you have to keep an eye on the newly-stocked dam anyway you will find that waving your arms and using the traditional cry of 'shoo' will work just as well. Many cormorants make a round of a number of dams during the day, and will arrive at approximately the same time each day. Take advantage of this by being there regularly when they arrive, and they will soon leave your dam off their visiting list. There will be inevitable visits later, when you are not present, but there is no point adding to the stresses of modern living worrying about this. When you initially stock your dam, your figures should allow for this natural loss rate.

There are several general remarks which should be made about keeping waterfowl on large bodies of water. Although they may well be able to forage for themselves under these conditions, they should regularly be fed small amounts by hand to prevent them from becoming too shy and difficult to observe. Like most native birds, waterfowl cannot be bought or sold without a special permit which can only be obtained if you can meet the standards set by the appropriate government body. Birds newly introduced to your dam will usually need their wings clipped to prevent them flying away, and should be confined in some way until they become used to their new home and owners. Foxes are the most serious threat to waterfowl; the birds will not necessarily sleep on an island where they would be safe even if one is present. Hence, provision must be made for a fox-proof area where they can be confined at night; possibly large sections of the shoreline will need to be enclosed. A fence of heavy chicken mesh (light mesh can be chewed through) 150 cm high has kept foxes away from my poultry for seven years even though these carnivores are very common where I live. The fence is supplemented with an overhanging panel 30 cm wide at the top, and a 60 cm width of the same mesh stapled to the ground at the base of the fence to discourage digging under. Such a fence can be effectively camouflaged by creepers, shrubs and trees.

# Other animals

Regardless of your plans, a number of small animals such as waterfleas, copepods, various bugs and beetles, etc. will make your dam or pond a home. Few of them present any problems, and many are fascinating to study with a 10× hand-lens which can be bought for just a few dollars at disposals stores. Populations of some insects may, on occasion, build up dramatically, only to disappear for years at a time. Others will maintain fairly steady populations over long periods of time. Almost none bite or sting with the exception of the bugs called Waterboatmen which have a sharp and irritating sting if handled carelessly (I must have handled literally millions caught in nets before I was first bitten), and several other bugs and beetles which may nip if you insist on picking them up. Most of these insects have a useful role in the ecology of the pond or dam — and many are directly or indirectly food for other animals you want to raise — adding a whole new dimension to watergardening which you should learn to appreciate, especially as you will never get rid of them anyway!

One animal which may move in of its own accord (even in large cities) is the native **water-rat**. This is not a disease-carrying slum-dweller like the introduced rats, but an animal well on the way to becoming something like an otter. They are shy and fascinating creatures to watch, and will help control potential pests such as burrowing freshwater crayfish. Their one fault is that they will burrow a home into the dam walls; however, this tendency can be kept in check if you provide shelter in the form of a few hollow logs in a corner of the dam where you don't mind the odd burrow. If you don't want them around, bear in mind that they are protected animals; your local National Parks people or equivalent authority will trap and remove them alive, if you want this done.

**Frogs** may also appear of their own accord, or you may want to introduce them deliberately by collecting frog spawn (eggs) and putting it in your pond or dam. Some people don't like having frogs because of their noisy calling, while others like them for that very same reason. If you have neighbours nearby, you may have to consider their feelings in this matter. Some frogs are particularly noisy; the winter chorus of bullfrogs in one dam on our property is so loud that you have to shout to be heard above it, though they do sound cheerful. You can identify many frogs simply by hearing their calls, and by their spawn. As the calling takes place at the same time of year as the eggs are laid, you can identify the frogs without actually seeing them, and the eggs are an easy way to introduce the ones you want without having to chase frogs all night!

**Tortoises** are another interesting addition to larger ponds or dams, but if you swim in the dam and like to dive in head-first, keep in mind that a large tortoise only differs from a rock in that it can appear unexpectedly wherever it wants to! They require no special care as they will feed happily on small fish and tadpoles.

# Problem solving

This last section of this chapter describes a miscellaneous assortment of problems you may come across, or which can be avoided.

## Turbidity

There are a number of causes for tinted water which I describe in the section on water quality. Some water, however, may be constantly cloudy or murky, and while it may often clear by itself, (especially in lined ponds), in some cases the water must be treated to make it reasonably clear.

Clay-coloured water often won't settle by itself, because the particles of clay may be so small that the slightest movement in the water will keep them afloat. Particles of this type can sometimes be made to bond together and form larger particles which will sink more readily, by scattering agricultural lime or gypsum over the surface of the water. Many waterplants help settle clay particles once they are established in sufficient quantities; the best example I know of such a plant is *Chara Fibrosa* (see under Australian Waterplants).

Grey or brown cloudiness usually occurs only in smaller pools and ponds and is associated with a build-up of bacteria due to an excess of decaying matter such as uneaten food or a dead animal. Discolouration of this kind is often accompanied by an unpleasant odour, however faint. Greenish water, on the other hand, is caused by minute plants and plant-like creatures, and is generally a sign of healthy living conditions. However, if the water is *too* green — the soupy-green of a duckpond for instance — then this is a result of excessive amounts of nutrients in the water. The nutrients may be washing in from another source, be from decaying food or a dead animal, or be a sign that there are too many animals in the water. Whether bacteria or green water are the problem, the cure is the same: decide which of the causes is most likely and act on it — a partial water-change will also help speed recovery to a more balanced state.

## Stock

Animals, especially cattle, should be fenced out of dams. They cause massive erosion, speed the ageing and silting-up of the dam, stir up mud, eat waterplants, and their droppings may over-enrich the water and be a possible source of disease. If you don't want to exclude them completely, limit their access to one relatively small section and lay a path of coarse rubble right into the water at that place to reduce erosion. In limited amounts their droppings may benefit the underwater life, though the water may not be hygienic for swimming. A better idea is to put a trough below the dam wall and keep it topped up from the dam with the help of a float-valve.

## Snails

Although freshwater snails are often introduced to ponds and dams as 'scavengers', they really do little in the way of cleaning-up. However, many will eat waterplants and some, particularly watersnails from sheep-farming country, may be carriers of liver flukes (a serious parasite in human beings too). Don't introduce plants from such areas as they will often carry these snails, which are hard to get rid of without elaborate treatment.

## Freshwater crayfish

Australia is rich in species of freshwater crayfish, and many are excellent eating. However, some species (particularly yabbies, *Cherax destructor*) will eat plants if allowed to build up in large numbers. They can also cause water turbidity and structural problems in dam walls with their burrowing activities. This tunnelling is especially triggered by falling water-levels, so keep the dam topped up to a constant depth and you will reduce this problem.

If you wish to reduce a large population of yabbies, some fish will eat the smaller yabbies, thus reducing their numbers in the long term while trapping will remove most of the larger breeders which fish can't eat. Fish suitable for reducing yabby populations include *Tandanus* catfish, golden perch, brown trout (but not rainbow trout), and possibly silver perch.

## Weeds

Unwanted plants are best controlled by establishing dense growths of the waterplants you *do* want to grow, as small or seedling weeds will have trouble establishing themselves among beds of healthy vegetation. Even then, it is worth going around larger ponds or dams once or twice in the growing season and pulling out anything undesirable — this is easily done while weedy waterplants are small. Like any garden, a watergarden does require a small amount of maintenance, but fortunately this is less (and more easily done) than for other forms of gardening. Weeding a dam is the perfect job for those stinking-hot summer days when it is impossible to work otherwise, as it is a great excuse for swimming around the dam and looking at everything in it!

## Evaporation and transpiration

Waterplants do not necessarily reduce the amount of evaporation from a large area of water. In fact, many transpire water (the plant equivalent of sweating) more rapidly than it would evaporate. However, there is a trade-off which requires control of plant numbers. A good example is cumbungi (*Typha* spp.); these can transpire water several times faster than it would simply evaporate. However, by planting strategically placed patches across particularly windy and exposed parts of a dam, the cumbungi can be used to break up dry winds, reducing the loss of water overall. Other species such as *Bolboschoenus* may do as good a job without being as difficult to control as *Typha.*

# Chapter Two

# Plants and Planting

It isn't always easy deciding which plants are waterplants and which really belong on drier land. This book treats waterplants as those which are naturally associated with water, and are adapted to spend at least a part of their life-cycle in fairly wet conditions. However, some plants are commonly grown in very wet conditions although they may not grow this way naturally, and in order to be thorough these have been included also. Before looking at growing conditions for waterplants and wetland plants, it is worth mentioning that many submerged plants are promoted as 'oxygenating' varieties which supposedly increase the amount of oxygen available in water; this myth is discussed in Chapter 1.

Probably the most important part of growing any plant described is knowing just how much water it likes. In general, deeper-water plants are not fussy about exact depths as long as the water is clear enough to let through a reasonable amount of light. For example, most waterlilies prefer a depth of between 30 cm and 50 cm of water over the crown, but will usually tolerate drops in waterlevel to as little as 15 cm without being endangered. In a dam in which water levels fall to below the level of the crown, waterlilies will usually still survive as long as they are still wet, though they will stop flowering until conditions improve.

Temperature is a very important factor in deciding the correct planting depth for deeper-water plants; in cooler parts of southern Australia too deep is too cold for more tropical plants, while too shallow (especially in summer) may be too hot for some temperate-zone plants. Using waterlilies again as an example, tropical varieties will thrive even in very hot, quite shallow water, whereas most hardy varieties won't do so well under these conditions, and some (particularly red or white varieties) may be very stressed by extreme temperatures. (The terms 'hardy' and 'tropical' as applied to waterlilies are explained in Chapter 5 — Nymphaea.)

The depths of water required by shallow-water plants vary, but most are perfectly happy with anything from 1–10 cm of water above the soil in which they are planted. Some varieties need to be in shallower or deeper water, and this information is included with the plant descriptions. Generally speaking, most plants living naturally in shallow waters are quite tolerant of variations in temperature and depth; they will grow in waterlogged soil just as happily as when they are submerged, and many will also thrive even in soils which are only just wet. Shallow-water plants also tolerate flooding, sometimes for long periods of time and to great depths.

# *Waterlogged versus wet or moist*

Waterlogged soils are those which are completely saturated with water, including all the spaces which would usually contain air. The soil may collapse to some degree and oxygen is unable to penetrate into it, so it becomes *anaerobic* (contains no oxygen). Anaerobic soils have a different chemistry to those with some oxygen. It is harder for plant roots to extract most nutrients from them, and they suffocate the roots of most plants, so only specially-adapted plants can survive these conditions.

Most wetland plants have specialised ways of getting oxygen to the roots to prevent the roots from suffocating. Many plants which grow fully submerged in water don't even bother putting their roots far into the soil. They obtain water and nutrients directly through their leaves, and their roots are mainly used as an anchor.

There is a dramatic difference between waterlogged soils and those which are just wet. This is that most wet soils are not anaerobic — there is just enough drainage to allow air to move through the soil. While many plants which grow in waterlogged soils will be perfectly happy in wet soils, the reverse isn't often true. Plants which like their soil wet but not waterlogged should be planted *above* waterlevel, usually far enough above to ensure that the crown of the plant and a reasonable amount of root are above the waterlogged zone. Even if some roots penetrate waterlogged soil, the plant can decide for itself just how far they will go.

Somewhere above the wet zone the soil is just moist (or damp). It is impossible to give an accurate definition of the difference between moist and wet because one changes gradually into the other; I leave this to common sense as most readers will recognise the difference at a touch.

Tussock-planting is a way of growing some of the more fussy wet-loving plants which won't stand waterlogging. You may have noticed tussocks of vegetation standing above waterlevel in natural swamps; these are often occupied by a distinct group of plants which may not occur elsewhere in the swamp. Such plants (for example, some *Lilium* species and several Australian swamp daisies) prefer constant wetness, but will only tolerate waterlogging for short periods of time.

Artificial tussocks which are both pleasing and functional can be made with a small amount of good-quality soil to which larger amounts of peat, vermiculite and similar materials are added, all overlaid with live mosses and held together with plastic mesh. The moss will grow through the mesh in a short time creating a convincingly natural tussock which can be cut to allow plants to be placed in it. This is also a good way to grow swamp plants with specialised requirements, such as carnivorous plants (in which case the mix used would be mostly peat with some sand).

# Direct planting or pots?

At whatever depth you are planting, you will need to decide whether to plant directly or into a pot. The advantages of pots are that they can be moved around as required, the spread of faster-growing plants can be easily controlled, and pots can be lifted out for propagation or maintenance as needed without draining the pool or stirring up mud.

The bottom of each pot should be lined with hessian, newspaper or other relatively non-toxic material to prevent soil from flowing out. Plants are potted in the usual fashion, but the surface of the soil should be covered with at least 2 cm of propagating sand or fine river-gravel to reduce the rate at which nutrients leach into the water, and to prevent goldfish or catfish from digging in the soil. When lowering the pot into water, hold it upright until most of the air has escaped from the soil, otherwise the pot may tip over suddenly if there is a rush of escaping air bubbles.

The type of pot used is not particularly important, though squat ones are frequently used because they are more stable and allow a greater depth of water over the top of the pot. Open-meshed waterlily baskets are commonly used in the northern hemisphere, the theory behind these being that they allow greater circulation of water through the soil. However, they are expensive and don't actually seem to make any difference to growth, even for waterlilies.

Direct planting is more common in very large pools and dams, especially if you want plants to spread freely. Line the areas to be planted with a blanket of specially prepared soil, then cover this with about 5 cm of propagating sand or fine river gravel. You should prepare an area for direct planting before water is added to the pool or dam, so if you are planting an already established dam you will need to partly drain it first.

## Bog gardens

A common situation for direct planting is in bog gardens. These can be built at the side of a pond or be a feature in their own right. If they are designed as part of a pond, make sure that water from the pool can flow freely into it so that the bog garden is automatically watered as long as the pool is kept topped up. Pipes, channels or other breaks through the wall of the pool will allow free movement of water, but the bog garden should preferably be as waterproof as the pool itself, otherwise it will be a major source of water-loss.

If you wish to include a bog garden as part of a lined pond, the simplest method of construction is to build it inside the liner. Where you plan to have the bog garden, make the pool from 10 cm to 40 cm deep depending on the amount of soil you will use, which in turn depends on the plants you intend to grow. Build an un-mortared stone or brick wall between the edge of the planned bog area and the deeper part of the pond, and line the bog garden side of it with some soil-retaining material such as hessian or a similar non-toxic material. The bog area should be filled with soil before the pool is filled with water.

A bog garden doesn't have to be associated with a pool. The simplest way to construct an independent bog garden is to dig a hole 30–45 cm deep, line it with heavy-duty black plastic, then fill with a suitable planting mix. Ultraviolet light destroys this type of plastic, but buried underground a black plastic liner may last a lifetime. Any small holes in the plastic will not make much difference to water loss; in fact, if you plan to grow only plants which like wet but not waterlogged soils, you will need to punch holes through the liner for drainage.

## Planting pockets and shelves

In a dam, larger plants can be started off in large pots. Allow enough time for the roots to bind the contents of the pot together thoroughly, then empty the root-ball out at the place and depth selected. If you have used an unusually rich mix in the pot, the plant will remain well-fed until it is thoroughly established. This is particularly useful for heavy feeders, such as waterlilies, as it gives them a concentrated source of nutrients for several years. However, direct-planting in specially prepared planting pockets and shelves is the more usual method of establishing plants in dams.

Planting pockets are a good way to both contain a plant and to make sure that it has adequate nutrients. Make a shallow hollow in the side of the dam at a suitable depth for the plants you wish to grow, then fill with a good-quality soil mix. Cover the soil with propagating sand or fine river gravel at least 5 cm deep to reduce nutrient loss, bearing in mind that the depth you are planting at is measured from the top of the sand (or gravel) to the surface, not from the bottom of the pocket.

For extensive plantings around a dam, planting shelves can be run around the edge as needed. These are really just planting pockets which run as far along the dam walls as necessary, and can be at any depth or width depending on what you plan to grow and how steeply the dam wall slopes. Planting shelves are covered with soil and sand as described for planting pockets. All work (including construction) is most easily done when the dam is first constructed, or after partly draining an established dam. If you plan to use a thick layer of soil over the shelf, angle the base slightly (away from the drop-off to deeper water) as wet soil is heavy and a thick blanket may otherwise slide into deeper water.

I have heard recommendations that topsoil saved during dam construction should be uniformly spread across the entire dam in the final stages, including the deeper parts. What this is supposed to achieve, I don't know; far better to save the soil for places where you have a use for it. Few waterplants will grow very deep in a dam other than weedy species, and a bare clay surface underwater in areas where you don't want plants will help control weeds.

# Soils and potting mixes for underwater use

Soils for underwater use should be quite similar to what we call open, freely draining soils used on dry land. They should be moderately rich, imitating to some degree the nutrient-rich (but often nitrogen-poor) substrate in an ancient swamp. Materials used to make such mixes should mostly be natural organic substances or mineral in origin, and not include rubbish like polystyrene pieces, otherwise you will end up with a mass of dandruff-like floating particles on the water. No matter how the soil mix is to be used, it should be covered with a layer of propagating sand, fine river gravel, or similar material to reduce leaching of nutrients into the water. This layer will also hold down materials such as vermiculite or peat which may float before they are thoroughly waterlogged.

An open structure in soils for underwater use is more important to allow quick root penetration and spread rather than to try to oxygenate the soil — most waterplants will grow happily in anaerobic soils anyway. By using oxygen provided by the parts of the plant which are above the anaerobic conditions, the roots of many waterplants can extract nutrients from the anaerobic soil which might otherwise be unusable.

Nutrients are best supplied by those fertilisers referred to as 'organic' (in the sense of non-synthetic) for several good reasons. Firstly, synthetic fertilisers dissolve readily in water, which means that they will not keep waterplants fed for long periods of time. Even slower-release fertilisers such as Osmocote and Nutricote dissolve quite rapidly when permanently submerged (although some waterlily growers now use these, the results can be unpredictable). Secondly, under anaerobic conditions the rate at which various elements of synthetic fertilisers dissolve may vary, and so they may not provide a balanced supply of nutrients. Thirdly, the nutrients in synthetic fertilisers are present mainly as salts. Many freshwater plants come from environments virtually free of salts, and don't respond well to their presence. Finally, many freshwater plants don't use the full range of nutrients provided in synthetic fertilisers as they are perfectly adapted to extracting their requirements from soils which have a completely different balance and availability of nutrients to terrestrial soils. In such cases surplus nutrients usually just leach out into the water where they can cause algal blooms and similar pollution-related problems.

Rather than describe a perfect soil mix for underwater use, I will discuss various materials and their use in such mixes; the exact proportions you use should depend on availability, the nature of the soil you start with and the requirements of the waterplants you intend to grow. At the most basic level, any cheap, organic potting mix enriched with a handful of blood-and-bone (which breaks down over a long period of time in anaerobic soils, making nitrogen available very gradually) for each 20 litres of soil will do. Heavy feeders such as waterlilies are best with up to 30 per cent cow manure by volume, and twice or even three times the amount of blood-and-bone used in mixes for plants with average needs for nutrients. Avoid the more expensive potting mixes unless you are certain that they don't contain dolomite or lime, which many bogplants dislike.

All necessary nutrients for most waterplants can be satisfactorily provided by a combination of *well-rotted* cow manure (fresher manure may produce bubbles in the soil which dislodge the plant) and blood-and-bone. Cow manure adds organic matter in a form which quickly changes to something similar to what is found in the mud of an old swamp, and it also provides some nutrients. It can be added at between approximately 5 per cent and 30 per cent of a mix depending on how quickly you want plants to grow and how heavily they feed. Blood-and-bone is a slow-release source of nitrogen and other nutrients scarce in cow manure; under anaerobic conditions, it may continue releasing nutrients for several years. Blood-and-bone can be added as handfuls or cupfuls at an approximate rate of 1 cupful to between 10 and 50 litres of mix, depending again on how rich a mix you would like to achieve. Other animal by-products such as poultry manure should only be used well-rotted and in very small amounts, or as part of a compost which should not exceed 10 per cent of the volume of the mix.

Peat, vermiculite, perlite and propagating sand (or fine river gravel) are used to develop soil structure to encourage root penetration, while keeping the soil heavy enough so that it doesn't become too slushy. An excess of sand will cause the mix to collapse under its own weight, but up to 20 per cent by volume (depending on how much sandy material is present in the soil you start with) will keep the mix open, yet fairly solid. Vermiculite and perlite are expensive but can be used in any amount you can afford as they are chemically fairly inert and very light. Vermiculite is always my first choice as it is similarly priced to perlite but will hold and absorb water well. For this reason it is particularly good for soils which at times may be above the water's surface, where it holds moisture much longer than perlite.

Peat, on the other hand, is hard to wet when dry, and is not good in mixes which may dry out as it is difficult to get the peat to absorb water again. Some types of peat may also be either too acid or too saline for most waterplants.

The main ingredient of any underwater mix is topsoil (which usually doesn't need to be sterilised if it is for underwater use as any weed seeds it contains will drown). The other materials described above are added mainly to adjust the properties of the soil you start with. For example, a poor and sandy soil would require a higher proportion of cow manure and blood-and-bone, while a rich, clay soil would need less in the way of nutrients, but more vermiculite, sand or peat to open up the structure. For the majority of waterplants, a pH range of between 5 and 6.5 is suitable as aquatic plants generally come from relatively soft-water environments which are moderately acid to neutral. There are exceptions, but these are included under the appropriate entries in Chapters 3, 4 and 5.

# Propagating waterplants

Most waterplants and wetland plants are easy to propagate by division or from cuttings, but a number must be grown from seed. While waterplants raised by division are as simple to produce in this way as any terrestrial plant, cuttings are often faster and produce more new plants with greater vigour, perhaps because many waterplants are adapted to re-establish themselves from pieces broken off in floods.

Only a very few named varieties or forms can be raised from seed — these are almost all wild plants which were taken from distinct populations and breed true if not cross-pollinated; others are sports which may throw a small proportion of plants similar to the parent plant from seed. A much higher proportion of species can be grown from seed, but the freshness of the seed itself is often the most important factor for good results. Some waterplant seed germinates quickly in the wild, and if allowed to dry out or prevented from sprouting will usually die. Other types of seed have a longer germination period and may survive years in a dormant state — the most extreme example being the seeds of *Nelumbo,* which may sprout more readily after 100 years or so of dormancy! In all cases the entries for individual plants (Chapters 3, 4 and 5) include such information as far as it is presently known.

## Buying waterplants

Waterplant prices in Australia are low by world standards — a grower in Queensland wholesales waterlilies to overseas markets for higher prices than the retail price within Australia! Few general nurseries keep waterplants or know anything much about them, although pre-packaged waterlilies and waterplants are sometimes sold among the bulbs. These packages are the worst possible way to start as they are often second-rate, unnamed varieties, and many will die from the rough treatment they have sustained in getting to the shelves of the nursery. While such plants are not expensive, better established plants of good, named varieties from specialist suppliers don't cost much more and will also flower much sooner.

There are a number of watergarden nurseries in Australia, but they are not universally good; some have been known to give incorrect or dubious information. One well-known watergarden has even sold non-native plants and goldfish as native species, and has apparently refused to alter the names of incorrectly named plants.

However, there are also a few, smaller watergarden nurseries which offer a reasonable range and prices, and helpful information. If you want to buy established potted plants, I would suggest that you buy from these smaller nurseries whose owners handle most sales personally and actually know something about aquatic plants. Even then, some misidentified plants have been around for a long time and may now be widespread under their aliases; examples include *Acorus calamus* and *Carex pendula* which are often sold as natives (the latter even under the name of a native, *C. fascicularis*).

If you don't live near a major city with watergarden nurseries, you can still buy virtually any waterplant described in this book by mail-order regardless of where you live in Australia. Waterplants are among the easiest plants to send by mail because many are adapted to long periods of flood, which is not all that different to what they experience when packed and sent by post.

# *Diseases, problems and pests*

I have little experience with the pests and diseases which affect waterplants. In part, this may be because Australia is free of many of the pests and diseases recorded overseas — for example, crown-rot in waterlilies is not known here and hopefully never will be if all new imports are properly quarantined before their release. Other problems simply don't often occur in watergardens — for example, most waterplants have developed a fair immunity to fungal diseases, apparently because they grow in conditions ideal for the spread of such diseases.

Watergardens are the one type of gardening in which a form of polyculture (growing plants and animals together) is routinely practised. Even the smallest pond usually boasts a couple of goldfish which will eat many of the insects and their larvae which might otherwise attack the plants. When pests of this kind *do* strike, however, it is impossible to treat them with any of the range of chemicals freely used in conventional gardening; nearly all of these will kill fish and other water animals very quickly, and most are persistent and require that the pond be drained and refilled several times before it becomes habitable again.

The most common insect pest is the humble aphid which can build up in plague proportions. Control is simple: blast them off the leaves with a jet of water and the fish will eat them. You may have to repeat the process several times, but the method works, is safe and is 100 per cent environmentally friendly.

Other insect pests can be controlled by the addition of a few tadpoles to the water; these will later develop into frogs that will leave the wetter areas around the pond and actively hunt pests from dusk till dawn in warmer weather. In fact, if your pond is large enough to support a reasonable population, you will find that they will control many other insect pests in the rest of your garden as well. If you are planning to try this method you must not use sprays anywhere, as frogs are perhaps even more sensitive to these than fish — it may be that many species in the wild are endangered partly because of this.

A final and very important factor which has contributed to the relatively problem-free watergardening I have generally enjoyed is that I use the underwater version of organic methods in my nursery. I have discussed the simulation of natural, underwater soils earlier in this chapter, and also why synthetic fertilisers should not be used; it is these semi-natural soils that are the key to success with waterplants. A specialised plant growing in soil conditions close to those in which it naturally grows will usually be more trouble-free than one forced to grow in alien conditions. A noticeable proportion of waterplants I have experimented on with synthetic fertilisers in the past have looked unhealthy from the start; some have been prone to insect attack, quite unlike otherwise identical plants grown in semi-natural soils.

In summary, most waterplants and wetland plants are as easy to grow as the easiest of garden plants, providing that they are grown in soil and water conditions not too different from those to which they are adapted. In most cases, this means soft or only moderately hard water, soils largely free of salts in their simple and easily ionised forms, a plentiful supply of organic matter simulating some of the conditions found in ageing swamps, nitrogen present only in a form which becomes available gradually, and plenty of sunlight. There are exceptions of course, but these are mentioned under the individual entries in the two chapters which follow, as are other problems or disease-like conditions which only affect particular plants.

# Chapter Three

# Australian Waterplants

## Introduction

Until fairly recently only a handful of native Australian waterplants were available commercially, and there is still much to be learned about the cultivation of a number of species. Those which can be grown in well-watered garden beds as well as really wet or waterlogged soils are better known, and selected forms of these have been grown for years. With time, more selected forms of various aquatics will also become available, especially as interest in them as an alternative to the weedier non-natives grows. It is worth commenting that some native waterplants probably have considerable weed or nuisance potential themselves; this information is included in the text where it is known.

Much of the new information included on cultivation is from my own nursery, but there are many Australian waterplants with horticultural potential which have not yet been cultivated successfully. Even the naming of dozens of the well-known aquatics has been in a state of flux in recent years as botanists sort out some of the major nomenclatural problems. In most cases, the names used in this chapter are currently regarded as correct, while some minor exceptions are explained in the text.

The various degrees of wetness tolerated by various types of waterplants is discussed in Chapter 2; it is essential that the reader read that section before this chapter. While the distinction between moist, wet and very wet conditions is subjective to some degree, the difference between wet and waterlogged conditions is clearly defined and must be understood.

### Nomenclature

All plants in this chapter (and also in the next one) are listed in alphabetical order using their scientific names, also known as their binomial names as they consist of two words. The first name is the *genus* (plural: *genera*) the plant belongs to; this is a group of very closely related species which share that same first name. There may be only a single species in a genus, or there may be hundreds. The second name is the *species* name, and the combination of genus and species is the binomial name. Although a number of closely related species can share the same genus, only one species in each genus can have the given species name, and so the combination of the *generic* and *specific* names is unique to one species of plant. This system is used to guarantee that every binomial name refers precisely to one plant only (common names are much less precise).

Each entry which follows is the name of a genus. Following that, the water-loving species, varieties and hybrids known to be cultivated from that genus in Australia are described with any appropriate details. Where large numbers of aquatic and moisture-loving plants belong in the one genus, an account is given of the most interesting, attractive or useful forms grown. Where there are so many that it is hard to single out the best, only a general account is given. After the initial entry, the genus is abbreviated to just the first letter of the generic name. For example, after the *Carex* entry *C. fascicularis* refers to *Carex fascicularis.* The name in parentheses at the end of each entry is the botanical family name to which that genus belongs. Synonyms are only mentioned if name changes are either recent or not well known.

There are also descriptions of planting situations (waterlogged, wet, etc.) before individual entries. These don't include situations in which plants will not thrive; they are intended only as a guide to help the reader find those plants which are well-suited to a particular location or situation in or around any watergarden. There may be some plants in the genera shown that will also grow beyond the range of conditions under which they are noted; however, other than in times of flood or drought they are generally naturally limited to the conditions described.

# *Australian plants suited to various uses and conditions*

**Australian wetland plants tolerant of both winter flood and summer drought conditions** can be found in the genera listed below. Although these species may survive extreme drought in some cases once well-established, they won't do well under such conditions and are therefore not necessarily listed as being well suited for drier conditions.

*Baumea, Bolboschoenus, Calocephalus, Carex, Chorizandra, Eleocharis, Empodisma, Gahnia, Glyceria, Gratiola, Helichrysum, Isolepis, Juncus, Lomandra, Marsilea, Muehlenbeckia, Myriophyllum, Patersonia, Phragmites, Restio, Triglochin, Typha, Villarsia.*

**Edible Australian wetland plants** can be found in the following genera:

*Aponogeton, Bolboschoenus, Brasenia, Eleocharis, Lemna, Nelumbo, Ottelia, Phragmites, Schoenoplectus, Spirodela, Triglochin, Typha, Urtica, Wolffia.*

**Shade-tolerant Australian wetland plants** can be found in the following genera:

*Alisma, Azolla, Blechnum, Carex, Ceratophyllum, Crassula, Cyperus, Dianella, Dicksonia, Elatine,* Ferns, *Gratiola, Isotoma, Lilaeopsis, Lythrum, Ludwigia, Marsilea, Mentha, Myriophyllum, Persicaria, Philydrum, Potamogeton, Ranunculus, Riccia, Ricciocarpus, Rumex, Samolus, Todea, Typha, Utricularia, Vallisneria, Viola.*

**Australian plants suitable for deeper water** (usually over 20 cm) can be found in the following genera:

*Aponogeton, Baumea, Brasenia, Chara, Eleocharis, Glyceria, Hydrilla, Isolepis, Ludwigia, Marsilea, Myriophyllum, Nelumbo, Nitella, Nymphoides, Ottelia, Phragmites, Potamogeton, Schoenoplectus, Sparganium, Triglochin, Typha, Vallisneria.*

**Australian plants suitable for shallow water** can be found in the following genera:

*Alisma, Aponogeton, Baumea, Bolboschoenus, Brasenia, Callitriche, Carex, Crassula, Elatine, Eleocharis, Epilobium, Glyceria, Gratiola, Hydrilla, Isachne, Isolepis, Isotoma, Juncus, Lilaeopsis, Lythrum, Ludwigia, Marsilea, Mimulus, Monochoria, Montia, Myriophyllum, Nelumbo, Nitella, Nymphoides, Ottelia, Philydrum, Phragmites, Potamogeton, Ranunculus, Restio, Rumex, Samolus, Schoenoplectus, Sparganium, Triglochin, Typha, Vallisneria, Villarsia.*

**Australian plants suitable for waterlogged soils** can be found in the following genera:

*Alisma, Amphibromus, Baumea, Blechnum, Bolboschoenus, Callitriche, Calocephalus, Carex, Chorizandra, Crassula, Cyclosorus, Cyperus, Damasomium, Dicksonia, Elatine, Eleocharis, Epilobium, Glyceria, Gratiola, Helichrysum, Isachne, Isolepis, Isotoma, Juncus, Lilaeopsis, Lythrum, Ludwigia, Marsilea, Melaleuca, Mimulus, Monochoria, Montia, Myriophyllum, Nelumbo, Persicaria, Philydrum, Phragmites, Ranunculus, Restio, Rumex, Samolus, Schoenoplectus, Sparganium, Triglochin, Typha, Villarsia.*

**Australian plants suitable for wet soils** can be found in the following genera:

*Alisma, Alternanthera, Amphibromus, Artanema, Asperula, Bauera, Baumea, Blechnum, Bolboschoenus, Brachyscome, Callistemon, Callitriche, Calocephalus, Carex, Celmisia, Chorizandra, Cotula, Craspedia, Crassula, Crinum, Cyclosorus, Cyperus, Damasomium, Dianella, Dicksonia, Eleocharis, Empodisma, Epacris, Epilobium, Gahnia, Gleichenia, Glyceria, Goodenia, Gratiola, Gunnera, Helichrysum, Isachne, Isolepis, Isotoma, Juncus, Lilaeopsis, Lobelia, Lomandra, Lycopus, Lythrum, Ludwigia, Marsilea, Mazus, Melaleuca, Mentha, Mimulus, Montia, Muehlenbeckia, Nertera, Patersonia, Persicaria, Philydrum, Phragmites, Pratia, Ranunculus, Restio, Rumex, Samolus, Schoenoplectus,*

*Schoenus, Scleranthus, Selaginella, Sparganium, Sticherus, Todea, Triglochin, Typha, Urtica, Villarsia, Viola, Xyris.*

**Australian plants suitable for moist soils** can be found in the following genera:

*Alternanthera, Artanema, Asperula, Banksia, Bauera, Baumea, Blechnum, Brachyscome, Callistemon, Calocephalus, Carex, Celmisia, Chorizandra, Cotula, Craspedia, Crassula, Crinum, Cyperus, Dianella, Dicksonia, Eleocharis, Empodisma, Epacris, Gahnia, Gleichenia, Goodenia, Gratiola, Gunnera, Helichrysum, Isolepis, Isotoma, Juncus, Lobelia, Lomandra, Lycopus, Lythrum, Mazus, Melaleuca, Mentha, Muehlenbeckia, Nertera, Patersonia, Persicaria, Pratia, Restio, Schoenus, Scleranthus, Selaginella, Sticherus, Todea, Urtica, Viola, Xyris.*

**Australian plants which float** can be found in the following genera:

*Azolla, Ceratophyllum, Lemna, Riccia, Ricciocarpus, Spirodela, Utricularia, Wolffia.*

# *Alphabetical list of Australian waterplants and wetland plants*

*Acacia* Some *Acacia* species are supposed to tolerate wetness, but from what I've seen of them they don't like permanently wet conditions. These include *A. elongata, A. extensa, A. paradoxa* and *A. riceana.* Another species, *A. baueri* (which grows in areas with a deep, permanent watertable), will not survive waterlogging for long but does need constant moisture. (Mimosaceae)

*Ajuga* Only one species of this genus is generally available; this is Austral Bugle (*A. australis*) which forms a small shrub. The spikes of purple-blue flowers appear in spring, but other colour forms are sometimes available, including pink, blue and white. Bugle will grow in full sun and fairly wet conditions, but will not tolerate waterlogged soils for more than a few days. All selected colour forms are propagated by division. (Lamiaceae)

*Alisma* Water Plaintain (*A. plantago-aquatica*) is widely spread throughout the world in variable forms. This plant for shallow water and wet soils has striking foliage — broad, bold leaves at the end of upright stems which resemble some non-native *Echinodorus.* The tall flowerstems during summer are a bonus, with tiny, white flowers opening over several weeks. These decorate a lattice-like mass of tiny branches which dry and keep well once seed has set. *Alisma* spreads quickly by seed, so it may become a pest if introduced into areas where it can spread (fortunately seedlings up to a year old pull out easily). The plant may become dormant and disappear in very hot weather if exposed to full sun, and it also dies down in the coldest months. Watersnails seem to be the only pest, stripping the fleshy stems and leaves. Propagation is easy from reasonably fresh seed sown on waterlogged soils, or by division of older plants. (Alismataceae)

*Alisma plantago-aquatica*

**Alocasia** Cunjevoi (*A. macrorhizos*) is a subtropical, Taro-like member of the Arum family which occurs naturally in rainforests. It is prone to fungal attack and rot if grown in cold, wet conditions and should only be planted near ponds in warmer climates than those in most of southern Australia. (Araceae)

**Alternanthera** *A. denticulata* labours under the name of Lesser Joyweed (I have not encountered the Greater Joyweed myself — perhaps it is illegal?). The flowers are inconspicuous and the plant is grown as a 'groundcover', forming a low, tangled mat. Moist to wet soils and a sunny position are ideal, but it will also tolerate flooding and drying out for weeks at a time. Divisions and cuttings take root easily if kept quite wet. (Amaranthaceae)

**Amphibromus** Commonly known as Swamp Wallaby-Grasses, this genus has many species which, from a gardener's point of view, are rather hard to tell apart. Only two have been offered to my knowledge, both growing in shallow waters or on wet soils. Common Swamp Wallaby-Grass (*A. neesii*) is usually a low, sprawling grass which will tolerate dappled shade. Dark Swamp Wallaby-Grass (*A. recurvatus*) prefers full sun and is probably the showiest member of this genus, with small, purple-plumed flowerheads in late spring. Neither species seems fast-spreading even under ideal conditions. Propagate by division. (Poaceae)

***Anigozanthos*** Kangaroo Paws generally prefer well-drained soils, but *A. flavidus* and *A. viridis* will tolerate quite wet (but never water-logged) conditions. However, they cannot be regarded as wetland plants by any means and should be set a reasonable distance back from wetter soil if used near a pond or dam. (Haemodoraceae)

***Aponogeton*** Native *Aponogeton* species can easily give you the impression that they are very happy and growing well for the first year or two they are kept, but this bright start relies entirely on what is stored in the bulb; once the bulbs are exhausted, the plant wastes away rapidly. Intense light is essential for all species, especially if they are kept in aquaria. Only one native *Aponogeton* will grow without greenhouse conditions in southern Australia, and then only in reasonably warm parts.

*A. elongatus,* the Queensland Lace Plant, is most commonly offered in aquarium shops. Grown under aquarium conditions, the bulb wastes away over several seasons, but in a brightly lit location 30 cm or so deep this could be a permanent resident of the pond from Sydney northwards (dying back in the cooler months). Propagation from seed can be done if you don't want to deplete wild populations (which is where most bulbs are presently obtained), though it takes time to build up flowering-size bulbs. *A. bullosus* is also available occasionally through aquarium shops, but is much more sensitive to cold. This beautiful species with rich-green, bubbly looking leaves will linger painfully for several years in southern Australia. It really needs a minimum winter temperature of 15°C and a long growing season at much higher temperatures to establish itself. (Aponogetonaceae)

***Artanema*** *A. fimbriatum* is a small shrub with showy, blue flowers during the warmer months. It is frost-sensitive and should be treated as an annual in cold areas, but may put out new growth if the frosts have only been light. Although it likes wet conditions it will not tolerate waterlogging for long. Propagation is usually by seed. (Scrophulariaceae)

***Asperula*** Woodruffs are small, bushy, slow-spreading plants with fine, attractive foliage and very small flowers. *A. gunnii* is a subalpine species which doesn't do well in warmer areas, while *A. subsimplex* is more tolerant of heat. Other species may be available occasionally, but few are real wetland plants in any sense. All *Asperula* can be propagated by cuttings or division of self-layered plants; those associated with water are reasonably fast-spreading in permanently moist or wet areas, but are easily swamped out by more vigorous plants. (Rubiaceae)

**Azolla** Both Australian species are fast-growing, floating ferns. On large areas of water, they may form dense carpets which cut off oxygen to the water below, and they can die off at times, rotting and fouling the water. However, they are a very decorative plants with their tiny, waxy and neatly arranged leaves, changing colour from green to various shades of red and pink according to sunlight and weather. If you don't mind netting out surplus plants regularly in their growing season they are an attractive addition to a pond or dam, but the larger the area the more work is involved in regular removal.

*Azolla filiculoides* (left), *A. pinnata* (right)

All *Azolla* species contain blue-green bacteria which fix nitrogen, so they are a useful source of green manure which can be dug directly into soil if necessary. They make a good-quality, weed-free mulch as well, but break down rather quickly. In some countries they are also used as a protein-rich dietary supplement for poultry and pigs. *A. pinnata* forms triangular plantlets which grow mostly in warmer weather and are most often brick-red. The plantlets of *A. filiculoides* are more like a snowflake in pattern and grow under cooler conditions, changing to a rusty-pink colour on occasion. Both species may disappear in some seasons, but will reappear when the weather is more to their liking. (Azollaceae)

**Banksia** *B. oblongifolia* and *B. occidentalis* may occur in swampy areas but are more likely to survive in reasonably moist soils rather than wetter conditions. The Swamp Banksia *B. robur* is less sensitive to wetter conditions and colder climates, and will grow in moderately wet soils. As this species has a fairly hefty root system (and wet soils are often waterlogged not far below), it is most likely to succeed planted well back from the water's edge in permanently moist soil. Full sun or only light shade is necessary; propagation by seed is simplest, but cuttings are also possible to strike if you are reasonably skilled with the methods for this genus. (Proteaceae)

**Bauera** River rose (*B. rubioides*) is a bushy, fine-leaved shrub with small, rose-like flowers which may be white through to deep pink, or be fuzzily striped in white and pink; forms with double petals also occur. The plant itself may reach 2 m in height and width; it is taller and stragglier in shade and more compact in full sun — prostrate forms are also grown. Although all forms are associated with river banks they will not tolerate prolonged flooding, and grow best on permanently moist to wet soils. Propagation of named forms is always by cuttings. (Cunoniaceae)

**Baumea** A fairly large genus with considerable variation in size and habit, all are fairly upright reeds which do not spread as rapidly or uncontrollably as some other sedges. Bamboo reed (*B. articulata*) grows to 2 m or more in height, forming dense stands in water up to 40 cm deep. The dark-green stems look like some types of bamboo cane and are graced by 30 cm-long, tawny-copper flowerheads which weep gracefully and remain attractive for months. The form of Bare twigrush (*B. juncea*) in cultivation is usually a wiry, very upright species not much over half a metre high with attractive waxy blue-grey stems; other forms may be taller and less well-coloured. *B. rubiginosa* is similar in effect to *B. juncea* yet is not as upright or as strikingly coloured. However, the flowerheads of this species are more conspicuous due to their rusty colour. *B. tetragona* is another medium-height species with deep ridges running along the length of each dark-green, rough-textured stem. The four-sided stems are unmistakable, but I have also recently found an apparently miniature form of the same plant — what this does in cultivation remains to be seen. Most species need sun for good growth but will accept dappled shade, and are also very drought-tolerant once established. Propagation is usually by division as not all species seem to set much fertile seed. (Cyperaceae)

**Blechnum** Not all Water Ferns are associated with standing or running water, but some prefer really wet conditions. Of these the ones described below have fishbone-type fronds, although other variations on this basic leaf-type are available through ferneries on occasion. *B. fluviatile* is particularly fast-growing, and requires shaded, fairly wet conditions. *B. minus* will grow in full sun if kept wet to waterlogged, as will *B. indicum* which will even tolerate slightly brackish water. *B. nudum* is a riverbank species which prefers less moisture under normal conditions but will tolerate flooding for weeks at a time (older specimens develop a black trunk which may reach a metre in height). *B. penna-marina* is a miniature, alpine species for wet and shaded areas forming dense carpets of fronds rarely exceeding 25 cm in warmer gardens. *B. wattsii* will grow quite well in permanently wet soils, but also occurs in drier areas than the other species. Propagation is from spores, or by division. (Blechnaceae)

*Bolboschoenus* Clubrushes are a group of very similar-looking, upright-stemmed sedges with long, slender, weeping leaves forming dense stands which give a bamboo-like effect. This genus was formerly included in *Scirpus*; all species make excellent shelter for waterbirds, but die back to large, edible tubers in winter (the younger tubers are softer and can actually be chewed — a great advantage in a plant grown for food). Best growth is achieved in full sun or dappled shade, and in shallow water or waterlogged soils, but they are also quite drought-hardy once established. River bamboo (*B. medianus*) will form thickets over 2 m tall, while Sea Clubrush (*B. caldwelli*) is only half that height and not as leafy-looking. Suckers will take root readily in warmer weather, but dormant tubers can also be transplanted. (Cyperaceae)

*Bolboschoenus medianus*

---

*Brachyscome* Most species are small- to medium-sized perennial herbs which may become untidy looking if not pruned but produce masses of beautiful (if small) daisy flowers in various shades of white, mauve, pink and blue. Those listed below like quite wet conditions, but do not like waterlogging. Full sun gives the best results as shading will affect flowering. There are too many very similar-looking species to warrant individual descriptions, so I have listed only those which may be available from nurseries and seed suppliers. These are: *B. acuminata, B. angustifolia, B. cardiocarpa, B. diversifolia, B. iberidifolia, B. multifida, B. nivalis, B. scapigera, B. segmentosa,* and *B. spathulata.* (Asteraceae)

**Brasenia** Water Shield (*B. schreberi*) is the only species in this genus, and extends across Asia, into Europe, and also North America. Related to the waterlilies (and more closely to the non-native *Cabomba*), it will grow from centimetres deep to a couple of metres if allowed to creep into deeper water by itself. Full sun or semi-shade makes little difference to the appearance of the plant, but it dies back in winter in colder climates and may do so even under warmer conditions if there is not enough sunlight. Once the stems reach the water's surface, they form a dense, flat carpet of oval leaves resembling shields, with the stem joining the centre of the leaf. The purplish flowers are waterlily-like and quite small — not showy but fascinating nevertheless. All young underwater growth is covered with a thick jelly which is not unpleasant to touch and seems to protect the plant from watersnails.

This plant has disappeared from many parts of its original range, especially in southern Australia, but may be recovering in some areas. Divisions establish most quickly in warmer weather, but even dormant pieces will transplant providing they have a fair amount of established root; in either case *Brasenia* is slow to establish. (Cabombaceae)

**Callistemon** Bottlebrushes need no introduction to native plant enthusiasts. As their cultivation is covered in almost every book on Australian plants I have limited my list to a number of species which are naturally associated with swamps and rivers and do well if kept permanently wet. However, they should be used with some care in a native watergarden as they may give an unnatural appearance if treated as permanent waterplants — under natural conditions they would be flooded or very wet for only part of the year.

Species for moist conditions (these will tolerate really wet but not waterlogged conditions): *C. linearis, C. macropunctatus, C. montanus, C. pallidus, C. pinifolius, C. pityoides, C. salignus, C. sieberi, C. speciosus, C. subulatus,* and *C. viminalis.* Species tolerating waterlogged conditions (these will also tolerate flooded conditions for long periods of time, though whether such a prospect fills them with joy is a matter for conjecture): *C. citrinus, C. sieberi* and *C. paludosus.* (Myrtaceae)

**Callitriche** Native Starworts are rare these days, possibly because introduced Starworts (*C. stagnalis* and *C. hamulata*) have invaded just about every place where they once grew. *C. muelleri* is the only really conspicuous and aquatic species likely to be available, and has all the attractiveness of the introduced species without being weedy. This species forms light carpets of small, rounded leaves in sun-dappled shallows or deep shade, and is propagated by division. (Callitrichaceae)

**Calocephalus** *C. citrinus* is rather like a small *Craspedia globosa* (though with more elongated flowers) and can be cultivated in the same way. It is not a true bogplant, yet it will survive in standing water for weeks without apparent harm. (Asteraceae)

**Carex** This group of grassy-looking sedges may have unusual and attractive flowerheads. Unfortunately, some aren't easy to tell apart and a few misidentified plants have been sold under this name (including a *Cyperus* species!). The flowerheads of *C. gaudichaudiana* are slender and glossy black, each flower marked with a green border giving the appearance of the scales on a lizard's tail. The more usual forms available may reach around 60 cm in height and are a deep green, though a really distinctive, dusty blue-green form from Tasmania (Bruny Island) is only half the height. All forms will grow in sun or shade, but the Bruny Island form shows richest colour in full sun. In *C. fascicularis,* the flowerheads are soft yet spiny-looking and green, and hang like foxtails along their stems.

The first two species described like permanent moisture or shallow water and thrive in sun or shade, but others will also grow in drier summer conditions. These include *C. appressa* with compact tussocks of tough, wiry leaves, and long flowerstems which weep gracefully on larger plants. *C. bichenoviana* is extremely tolerant of dry conditions; it can look superficially similar to some forms of *C. gaudichaudiana,* but the blockier, darker flowerheads are quite distinct when quite new. *C. tereticaulis* is taller, less grassy-looking and more upright than other *Carex* grown and gives a natural look to wild plantings. A number of other, less

*Carex fascicularis*

striking, smaller species are available on occasion, but they are mostly plants of moist, forested areas rather than wetland plants as such. Propagation is usually by division, though some species set enough fertile seed to be worth growing on. (Cyperaceae)

**Celmisia** These low-growing alpine daisies form drifts of white flowers in the high country in summer, but can be short-lived at lower altitudes; cultivation is as for *Brachyscome* although some species may be more tolerant of waterlogging. Only *C. asteliifolia* is likely to be available with any regularity, however, *C. longifolia, C. saxifraga,* and *C. sericophylla* require similar conditions and would be worth trying when seed is available. (Asteraceae)

**Ceratophyllum** Hornwort or Foxtail (*C. demersum*) is a floating plant which is often offered for sale in aquarium shops, but the forms offered are not necessarily native ones and can vary immensely in their cultivation requirements. Hornwort provides good shelter for smaller fish near the surface. It may sink in cooler weather and will not necessarily reappear from very deep water. The rather coarsely textured rosettes of narrow leaves encircling the stems may drop off if the plant has been subjected to rough handling or dramatic temperature changes. In ponds where it is happy (usually reasonably shaded and fairly constant in temperature) it is a beautiful plant, but if you have trouble growing it, try some of the *Myriophyllum* species instead. A number of these have similar or even finer underwater foliage, yet are much less fussy in their growing requirements. (Ceratophyllaceae)

**Chara** Stoneworts are unusual algaes which form dense tangles of growth in moderately deep water. There are probably several species in Australia but nomenclature in this genus is not always certain, and they show variation under different conditions of light, temperature and planting depth. On the positive side, stoneworts form lush tangles (sometimes with contrasting red or orange fruits) which provide good shelter for smaller water animals such as shrimps. On the negative side, they grow like weeds and every now and then may die off for no apparent reason, fouling water and producing an odour which is unpleasant to most people (though not to me, but then I also collect carrion-scented flowers!). At least one species (*C. fibrosa*) is so finely divided and dense growing that it may help to settle the fine clay particles which are a common cause of murky water in dams. Such clay particles are kept afloat by the slightest water movement unless they are made to clump together to form larger particles which sink more rapidly. Some clays can be made to form larger particles by adding lime or gypsum to the water, but *C. fibrosa* is so thickly matted that it provides the very still conditions in its 'foliage' which allow the finest clay particles to settle. It is also an attractive and feathery plant and shows no sign of dying off abruptly after a time, unlike several other *Chara* species. All *Chara* species can be propagated by tearing off pieces and simply weighing them down where you want with a suitably sized pebble (I suppose this is technically 'by division'). (Characeae)

Three native submerged plants with finely divided leaves: (left to right) *Chara* sp., *Chara fibrosa, Ceratophyllum demersum*. Although none produce roots, *Ceratophyllum* is a flowering plant, while the other two are closely related and unusual algaes.

*Chorizandra* Several Bristle-rush species occur in Australia, but only one is regularly available commercially. This is the Echidna rush (*C. enodis,* also called Black Bristle-rush) with wiry pinkish-grey stems and flowerheads which look remarkably like tiny echidnas climbing up the stems. Flowering in this species can be erratic, as a clone which produces masses of flowers in one year may produce none a year later, while other forms in the same conditions produce a similar number of flowers each year. Cultivation is not difficult; even the poorest soils will do, and plants will live happily in shallow water or permanent drought year-round. For best foliage colour grow Echidna rush in full sun without much feeding. Propagate by division. (Cyperaceae)

*Cotula* Mountain Waterbutton (*C. filicula*) will grow in wet or permanently moist soils in cooler parts of Australia. In bright sun the plant forms a low-growing mat of finely divided leaves with small, greenish, button-like flowers over the warmer months. *C. australis* is another finely divided mat of leaves covered with greenish button flowers, and is best suited to wet soils rather than waterlogged conditions. Another species (*C. coronipifolia*) has been regarded as a native in the past, but is now known to be introduced. (Asteraceae)

**Craspedia** Billy-buttons are daisies from boggy areas which will also grow in drier conditions in cultivation. They grow best in permanently moist to wet conditions with full sun, but several drier months over summer will not harm the plant. The most commonly available species is Drumsticks (*C. globosa*) with bright-yellow, spherical flowerheads on tall stems in late spring. The flowerheads of this species dry well, keeping their shape and colour, though most forms available commercially have fairly small flowerheads (flowerheads of better variants may be close to the size of a golfball). *C. chrysantha* and *C. glauca* have smaller, less rounded flowers and prefer cooler conditions because they are basically plants of higher meadows. Two recently discovered (and unnamed) species appear to be even more aquatic, but their cultivation requirements are unknown. Propagation is usually from seed, but selected large-flowered forms must be grown from division of clumps or cuttings. (Asteraceae)

*Craspedia globosa*

---

**Crassula** *C. helmsii* (Swamp stonecrop) is the only one of the few Australian species of Crassula which is truly aquatic. Most *Crassulas* are intriguing succulent plants from southern Africa, so an aquatic relative is quite an exception. This fast-spreading species forms an attractive 15 cm-high mat of finely divided, bright-green leaves in shallow water or on any moist or wet soil. It is so dense-growing that most weed-seeds which sprout below it don't find enough light to grow, hence it is excellent for suppressing weeds around the water's edge. Easily propagated by divisions or cuttings. (Crassulaceae)

*Crassula helmsii*

**Crinum** Crinums include many of the few completely aquatic bulbs, with some Asian and African species growing and flowering in shallow water for their entire lives. Australian species associated with water will only tolerate flooding for short periods of time and should be allowed to dry out reasonably well over the summer months. Murray lily (*C. flaccidum*) is the most cold-tolerant, but even so does not grow or flower reliably in cooler climates. Its usual flower-colour is creamy-white, though a much rarer, yellow form has occasionally been available.

The more sub-tropical *C. pedunculatum* will only survive moderate frosts to –4°C. This latter species is also available in a variegated form which is even more sensitive to cold, flowers poorly, and could be convincingly impersonated by dribbling yellow paint on the wild form. Flowering is not reliable in cooler areas in any clone I have tested, and propagation is usually by division, except in warmer climates where fresh seed is reliably available. (Liliaceae)

**Cyclosorus** *C. interruptus* is a scrambling yet elegant, frost-tolerant, fishbone-type fern with large, coarse fronds. Waterlogged to moist soils and sunny conditions are best for proper growth, but moderate shade is also acceptable. Propagate by division. (Thelypteridaceae)

**Cyperus** Leaf Rushes produce bunched flowerheads in a rosette of leafy bracts carried on a stem rising above the foliage in most cases; the base of the flowerstem being surrounded by leaves which look like the bracts but are generally larger. Most species tolerate flooding for weeks or months at a time, but prefer waterlogged soils which may also dry out over the summer. All species in cultivation will grow in full sun if kept permanently wet, yet most will also grow in shady conditions and are propagated by division.

*C. lucidus* may reach well over 1.5 m when in flower; the glossy, dark-green leaves have sharp edges which should be handled with care. Despite this drawback, it is a striking plant forming dense thickets; the dark, coppery-brown flowerheads are very conspicuous and shine like polished metal in the sun. *C. exaltatus* can grow almost as large, though the leaves are a brighter green and the carroty-brown flowerheads are more compact; this species prefers wetter conditions than many of the others.

*C. gunnii* is drought-tolerant once established. Its slender and very tough leaves form semi-weeping, open tussocks to more than 1 m in height (the flowerheads are quite inconspicuous). *C. lhotskyanus* is very similar if slightly larger and with broader leaves.

A number of smaller species are easily cultivated, but only *C. gymnocaulos* has been available with any regularity. This leafless species forms low-growing thickets of rigid, upright stems topped by small flowerheads so arranged that they give an impression of a tiny explosion. Two forms of the very variable *C. vaginatus* have also been grown; the most distinctive one is a very densely packed miniature otherwise resembling a leafier version of the non-native *C. involucratus*. The other form is a tall, upright and quite leafless version of the same plant (the non-native *C. involucratus* is closely related to both *C. gymnocaulos* and *C. vaginatus*). *C. mirus* (a thread-leaved, very dwarf species) has also been offered on occasion, but it has been sold incorrectly as *Carex inversa*. (Cyperaceae)

**Damasomium** Starfruit (*D. minus*) is very similar to a miniature version of the related *Alisma plantago-aquatica* and can be cultivated in much the same way. The most obvious difference is the proportionately large flowerstems covered with spiky-looking, starry clusters of fruits; the leaves are also less tapered at the base, giving a more heart-shaped appearance. Although Starfruit is usually annual, it will self-seed freely. (Alismataceae)

**Dianella** Flax Lilies are grassy forest plants rather than wetland plants, but two broader leaved species grow in such wet conditions in winter and spring that they can be treated as semi-aquatic. However, Flax-lilies are usually found on freely draining slopes, so no matter how saturated the ground may become, the water is usually moving and anaerobic conditions don't occur. *D. tasmanica* is regularly available and grows in height to over 1 m. The pendulous masses of deep-blue flowers have bright-yellow centres and are followed by glossy, blue berries which vary in colour and size but can be very large and very richly coloured on selected forms. *D. caerulea* is smaller and the flower colour may vary from pale to darker blue, though its berries are similar in appearance to *D.*

*Damasomium minus*

*tasmanica*. Both species should be grown in shaded conditions; while full sun can be tolerated, plants will look rather yellow and unhealthy. Temporary waterlogging will cause little harm, but normal growing conditions should be between moist and wet. (Liliaceae)

**Dicksonia** Soft Treefern (*D. antarctica*) is the treefern most frequently sold in Australia; huge numbers are offered each year, and most seem to be planted only to die from lack of proper care or indifference. Unlike most other native treeferns, this species can regrow roots from the trunk if this is simply cut off, but the trunk must be kept wet and reasonably protected from hot sun while it is establishing.

Once established, Soft Treefern grows best in shade; plants kept in full sun make rather stunted specimens and must be kept really well-watered. This species will grow in waterlogged soils providing the water is moving and draining freely (as it would at a creekside, for instance, but should otherwise only be kept moist to wet. Propagation from spores is possible but very slow so most plants are simply cut down in the wild, and from what I have seen many have been poached despite a system of documentation which should protect them. (Dicksoniaceae)

**Diplarrena** Whether there are two species in this genus, or they are just forms of the one species, these are not wetland plants in nature, even though they will grow in permanently moist or wet soils quite happily. *D. moraea* has white, rather iris-like flowers with yellow centres marked with some purple. *D. latifolia* is usually smaller growing with flowers larger in proportion to the plant. While these are similar in appearance to *D. moraea,* they are much more boldly marked with purple. 'Amethyst Fairy' is a selected, larger-flowered form with even more showy markings and has recently been made available. Full sun or light shade are best for flowering, and propagation is usually by division, though seeds can be readily raised if fresh. (Iridaceae)

**Duckweeds** Most of the common species of duckweeds don't look very different from each other from a gardener's point-of-view, as all are small-leaved, floating plants which form dense, bright-green carpets across still waters. No special effort at propagation is necessary as they will grow in sun or shade providing there are adequate amounts of nutrients in their water. *Lemna* and *Spirodela* include the typical, larger leaved species with fine roots dangling below, while among these you may sometimes find the tiny, millimetre-wide plantlets of *Wolffia,* the smallest flowering plants in the world. *Wolffia* is so small that it doesn't have root hairs, each part of the plant being able to take in water and nutrients directly from the outside world. The flowers are a tenth of the length of the plant itself and are hidden in a tiny pocket in the upper surface of the leaf.

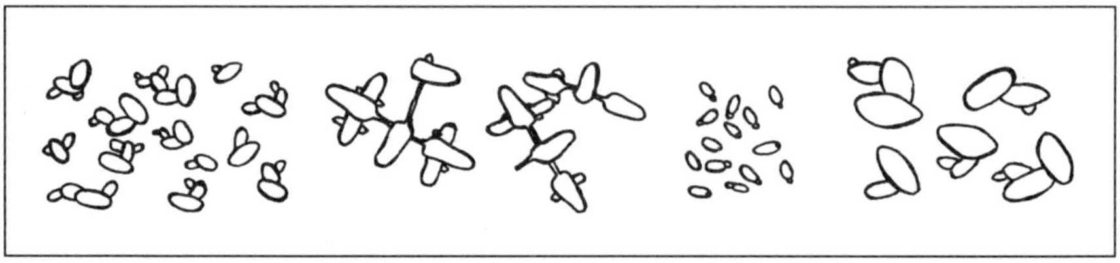

Duckweeds: (left to right) *Lemna* sp., *Lemna trisulca, Wolffia* sp., *Spirodela* sp.

Another distinctive duckweed is the Ivy-leaved duckweed (*Lemna trisulca*) which forms chains and tangles below the water's surface, and may even grow in moving water by hooking itself around other plants and continuing to grow from there. Dense growths of *L. trisulca* seem to inhibit growth of other duckweeds, starting by tangling their roots and physically preventing further spread. In windy conditions, thickets of *L. trisulca* may gradually tumble over, pulling under the competing surface-dwelling species and perhaps even drowning them.

Duckweed species are excellent fodder for some fishes, ducks and also plant-eating domestic animals. In the past, they have even been used as a fast-growing food source for humans. (Lemnaceae)

**Elatine** *E. gratioloides* is a tiny, creeping plant for shallow water or wet soils. Although not showy, it is an attractive bright-green and hides bare mud quickly under a dense, low carpet of tiny leaves, even in fairly shaded areas. Propagation is by division of stems or cuttings, or from seed. (Elatinaceae)

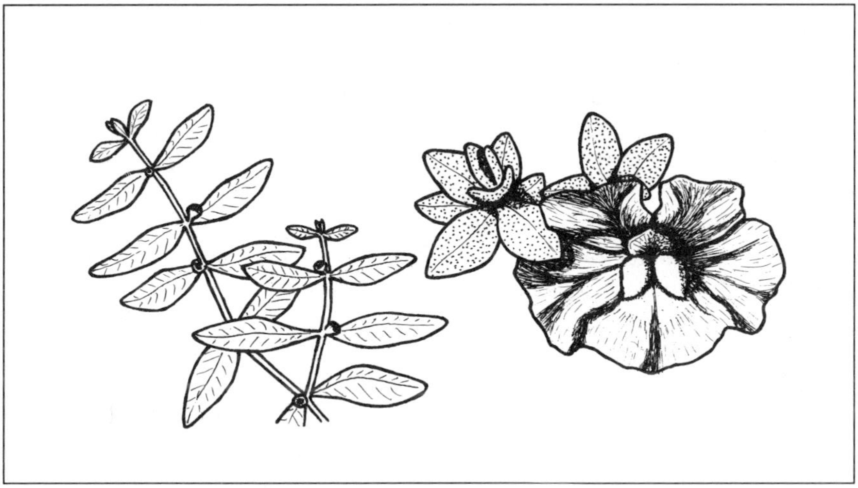

Two tiny-leafed native creepers for shallow water and wet soils: *Elatine gratioloides* (left) and *Mimulus repens* in flower (right).

**Eleocharis** Spikerushes vary enormously in size, but all are cylindrical-leaved, fairly upright rushes topped with cone-like flowerheads, and spread quickly if conditions are to their liking. *E. gracilis* is a small, semi-weeping species; my favourite form is a waxy blue-green clone which rarely exceeds 10 cm in height. *E. acuta* is similar but much more upright, green and at least twice the height.

Tall Spikerush (*E. sphacelata*) is a much taller plant (well over 1 m) which forms thickets of thick but hollow stems. Although it is very attractive, it can choke out smaller shallower dams in just a few years which makes it terrific waterfowl habitat if you don't care whether you can see any water. A native form of the Chinese Waterchestnut called Bulguru (*E. dulcis*, discussed under non-natives) can be grown in southern Australia but is neither ornamental nor productive. While all Spikerushes will grow in shallow water, they are also drought- and shade-tolerant (however, *E. sphacelata* grows best in deeper water); propagate by division or from fresh seed. (Cyperaceae)

**Empodisma** *E. minus* is a wavy-stemmed relative of *Restio* and looks best twining up through low, scrubby heathlands; grown alone they tend to become bushy and very tangled looking — attractive, but in a manic sort of way. Several distinct clones have been available from time to time, all of which will grow in full sun or with moderate shade in moist to wet locations. They also tolerate waterlogging or drought for considerable lengths of time. Propagation is by division or from fresh seed, but results are not reliable. (Restionaceae)

*Epacris* Many native heaths occur naturally in soils which can be very wet or even waterlogged in the wetter months, but none of those I have tested tolerate waterlogging for any length of time. All species mentioned below are adequately described in books that deal with conventional gardening with Australian plants, and are only mentioned here for the sake of completeness. Species which will grow reliably in moist to wet (but not waterlogged) conditions include *E. breviflora, E. exserta, E. lanuginosa, E. obtusifolia, E. pulchella* and *E. reclinata.*

Other species are reputed to tolerate waterlogged conditions or even shallow water indefinitely. However, I have tested a number of these and even short-term waterlogging may kill their roots very adequately. A simple test of their preferences is to stand a well-established potted plant in a shallow tray of water. Emptied out of its pot after several months of cold weather, there will be a clearly defined line below which the roots have perished; this is where the waterlogged zone begins. However, such species will tolerate waterlogging for a few weeks at a time, and include *E. microphylla, E. obtusifolia* and *E. paludosa*. (Epacridaceae)

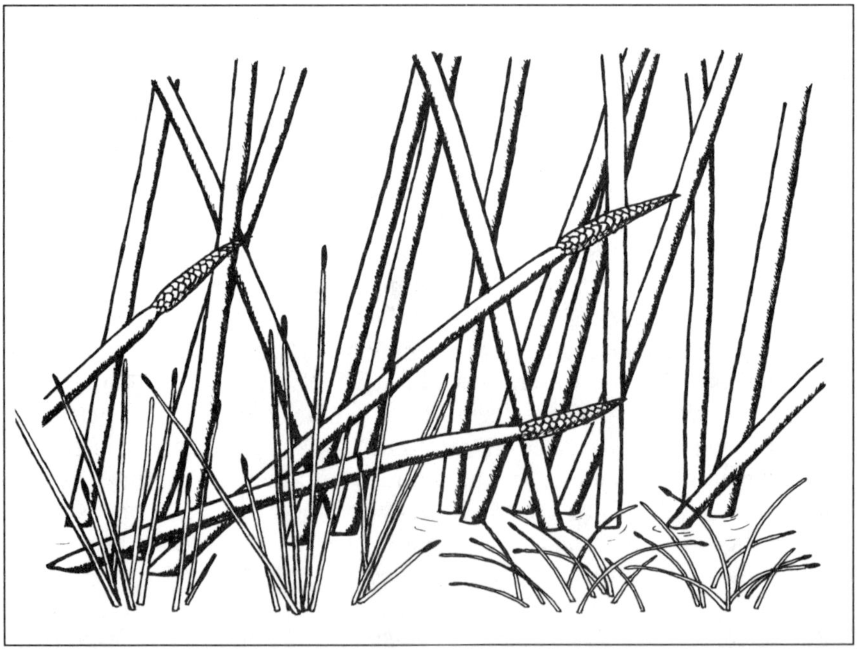

Native *Eleocharis* species: (background) *E. sphacelata; (foreground) E. acuta* (left) and *E. gracilis* (right).

*Baumea articulata* The large golden-brown flowerheads of this plant look as good as new for many months after forming. This is an excellent waterfowl habitat in dams where the invasive habits of *Typha* and *Phragmites* are not wanted.

---

*Brasenia scheberi* A relative of the true waterlilies, this unusual plant has purplish flowers nestling among the elliptical leaves. It has vanished from many of its old haunts in Australia, so growing it may contribute to its conservation.

*Chorizandra enodis* Although long available in native nurseries, Echidna rush is not seen as often as it should be. The flowers appear to be little echidnas climbing the pink-grey wiry stems of this very tolerant sedge which will grow in both drought-stricken summers or in shallow water for all of winter.

---

*Cyperus lucidus* is a big and bold grassy plant for wet soils, with rich green leaves and shining metallic copper flowerheads — handle the leaves carefully as they are sharp-edged. The dense thickets this species forms would make it a perfect hedge to keep people away from areas of a watergarden you don't want disturbed.

*Isotoma fluviatilis* is a variable plant which comes in all shades of lilac through to rich blue. The best forms (this very prostrate variety is one) form dense carpets of flowers which appear for months in any wet position. Taller growing forms will do well even in shallow water.

*Ludwigia peploides montevidensis* The bright and glossy yellow flowers alone are enough reason to grow this fast-spreading plant, but the fleshy leaves are also attractive, whether the plant is grown as a trailer over deeper water, or a low-growing herb for wet soils.

*Marsilea mutica* is an elegant native fern which grows vigorously in deeper water anywhere in Australia, and will do so even in countries where the water's surface freezes over for months.

*Myriophyllum papillosum* 'Red Dragonfly' and *Orontium aquaticum* are two of the finest waterplants in southern Australia. 'Red Dragonfly' is my own selection, and is probably the most richly coloured *Myriophyllum* in the world, with a deep crimson-red surface for most of spring. The bright and candle-like flowers of *Orontium* (a non-native waterplant) are seen just starting to appear above water and will soon be followed by blue-green waxy leaves which stand out of the water like lance-heads.

*Ottelia ovalifolia* is a truly lovely plant and the nearest thing in southern Australia to a native waterlily. The large fragile flowers appear for months among darkly tinted and heavily veined elongated leaves.

---

*Restio tetraphyllus meiostachyus* is perhaps the best-known of native grass-like plants. This soft and feathery rush is quite variable across its range; the form shown is lower growing and even more feathery than most.

*Caltha polypetala* has big rounded leaves up to 30 cm across, and giant buttercup flowers which make this plant quite unmistakable, but it looks its best only in cooler climates and even then needs shade.

*Cyperus papyrus* is probably the best-known of all sedges in cultivation in Australia. It prefers frost-free climates where it may form thickets up to 3 m high.

*Colocasia esculenta* All Taro varieties in Australia have leaves of much the same shape; rich colouring is an added attraction. The purely ornamental, purple-veined variety is the one most often seen, although some good edible varieties are grown here too.

---

*Hibiscus coccineus* (Louisiana Swamp Hibiscus) has a flower as big as my hand. The leaves are very ornamental and are similar in appearance to marijuana leaves, hence they often attract attention until the first flowers appear towards the end of summer.

*Iris* 'Gerald Darby' is probably a form or hybrid of *I. versicolor* and grows in shallow water or on wet soils. The flowers are freely produced on deep purple (almost black) stems over several weeks during early summer.

---

*Iris laevigata* 'Semperflorens' is close to the wild *I. laevigata* as you can get, with single-petalled flowers in the richest blue seen in irises, perhaps with a hint of purple. This form may flower continuously from late spring into autumn. The other colour forms of this species are no less attractive and are often very different from other irises in effect.

*Iris pseudacorus* forms and hybrids, from left to right: wild-type flowers, 'Sulphur Queen', 'Holden Clough' and 'Golden Queen' ('Superba').

*Lobelia* 'Queen Victoria' has quite surreal colours, almost beyond the boundary of good taste! The beetroot red leaves form a very contrasting background for scarlet columns of flowers which appear for months on well established clumps.

*Myosotis scorpoides* 'Mermaid' is a compact growing Forget-me-not for shallow water and wet soils. Flowers are both larger and bluer than the usual species naturalised in gardens.

---

*Nelumbo nucifera* This selected native form of the sacred lotus grows and flowers at lower temperatures than most other varieties. The flowers may reach 25 cm across; all parts of the plant are edible, and it will flower from January to April even in a cold summer — who could ask for more?

*Nymphaea* 'Chromatella' or 'Golden Cup' is the most commonly available yellow variety suited to smaller ponds. *N.* 'Moorei' is very similar in appearance but slightly deeper in colour, less vigorous and perhaps slightly less free-flowering.

*Nymphaea* 'James Brydon' is a compact plant which wouldn't be crowded growing in the space of the average-sized bath. The flowers are unusually rounded, deep pink and with a rather apple-like fragrance. This particular plant has over 30 flowers and mature flowerbuds, and is one of the most freely flowering varieties.

*Nymphaea* 'Loose' has beautiful and distinctive pure-white flowers which are very fragrant. Although this variety is supposed to hold the blossoms well above water, more usually they are seen floating in the classic waterlily position.

---

*Nymphaea* 'Rose Arey' If you have trouble selecting a rich pink waterlily from the dozens which are grown in Australia, try this one. The shape of the perfect, stellate flowers is complemented by their rich pink colouring (which may be paler in shallow water). This is a freely flowering variety with a strong fragrance reminiscent of aniseed and honey.

*Nymphaea stellata* is perhaps the best of the tropical blue waterlilies for southern Australia as it is cold-tolerant, fragrant and very beautiful. The colour varies between pale and medium sky-blue at different times of the growing season and in different climates; some forms may be pale pink.

---

*Nymphaea* 'Sunrise' has enormous blooms — often over 20 cm across — in a rich butter yellow. This is a large-growing plant which really only looks its best in larger pools, although it will grow fairly well in shallow water.

*Nymphaea* 'William Falconer' is probably the best deep-red fragrant waterlily for southern Australia, and is certainly my favourite. Other reds tend to be less even in colour or have too much white, weakening the redness of the inner petals.

---

*Nymphaea* 'W.B. Shaw' is regarded by a number of watergarden writers as the finest hardy waterlily. I would not want to have to make such a difficult choice, but it is certainly elegantly shaped, delicately tinted and one of the most deliciously fragrant flowers imaginable.

*Nymphoides crenata* 'Variegata' This widely distributed and variable plant has generally only been available in a small-leaved and rather amorphous-looking variety. Other forms are more mosaic-edged, and this cold-sensitive variety (which dies down in winter in southern Australia) is also patterned in purple-red.

---

*Sarracenia leucophylla* All pitcher plants are very eye-catching but this one with red to deep green marbling on a bone-white background draws the most attention of all. The rather futuristic-looking flowers are a pastel crimson colour.

*Scirpus (Schoenoplectus)* 'Zebrinus' is one of the most distinctive of the variegated grassy plants, with regularly banded stems resembling porcupine quills. Grown in full summer sun their colour becomes washed out, so keep them shaded at the hottest times of day.

*Zizania aquatica* This is a large-growing and striking annual grass with masses of delicate pink and yellow flowers, which is often grown for ornament in the northern hemisphere. Once the seed ripens, smoke it over a slow fire and it becomes that expensive delicacy, wild rice.

**Epilobium** The Willowherbs include many weeds which spread freely by seed, and can become impossible to control. In most cases, their flowers are small and very variable, so identification requires fruits and flowers. Only one species has really showy flowers — astonishingly, this is called the Showy Willowherb (*E. pallidiflorum*). Although the inconspicuous willowy-leaved plant of this species can easily be confused with many other *Epilobium* species, when it is in flower the lilac or purplish flowers will be visible from the other side of a dam. Although Showy Willowherb is a bit on the weedy side in cultivation, it will never swamp out other plants and has become quite rare in the wild so that it is listed as an endangered species in Victoria. Plants can be propagated by seed, division or cuttings, and will grow in full sun or semi-shade so long as they are always kept at least wet. They will also grow happily in shallow water or among tangles of other plants in deeper water. (Onagraceae)

**Ferns** Hundreds of native ferns require permanently moist or moderately wet soils, but don't like much wetter conditions. This book only includes those which are naturally associated with wetlands and which do best in wet, waterlogged or submerged conditions. The genera included in this chapter are *Azolla, Blechnum, Cyclosorus, Dicksonia, Gleichenia, Marsilea, Sticherus* and *Todea.*

**Gahnia** Sawsedges are usually large tussocks with sharp-edged leaves. Although occurring naturally in moist to wet locations, none will tolerate waterlogging for any length of time. Full sun or shade are acceptable, and propagation is often by division; however, many forms (particularly of the most frequently available species, *G. sieberiana*) must have a large amount of their immense root-systems dug out if they are to re-establish, and it is easier growing such plants from seed. (Cyperaceae)

**Gleichenia** Coral ferns form tall tangles of very-branched, beaded fronds and are a particular favourite of mine. For many years I had trouble establishing or transplanting plants of any species until I was told that they are best grown up from spores. Shortly after learning this, we bought our present property and discovered that we owned nearly an acre of Coral fern on a north-facing slope watered by permanent springs. Although I haven't tried to grow Coral fern since that time, new plants appear constantly around dams we have put in and are surprisingly fast-growing. The cost of *Gleichenia* species will always remain high because of the difficulties and the time factor involved in raising or establishing plants, but two species (*G. dicarpa* and *G. microphylla*) are often available through fern nurseries which propagate their own plants. Constant wetness and sun or semi-shade will allow for good growth — waterlogging or running water may be tolerated for months at a time. All species are best obtained as well-established plants in pots. (Gleicheniaceae)

*Gleichenia*

**Glyceria** Australian Sweetgrass (*G. australis*) is an upright, slender and elegant grass for wet soils or even quite deep water providing this doesn't remain too deep in warmer weather. This species is a good component for waterfowl habitats (providing some shelter and, indirectly, feed) and is fast-spreading without becoming so dense that it will choke out other plants. Dead leaves and stems have a tendency to accumulate in patches of live growth but are easily raked out with the fingers. Australian Sweetgrass is also reasonably drought- and shade-tolerant. Propagation is usually by division. (Poaceae)

**Goodenia** Goodenias are familiar to most gardeners with an interest in Australian plants; most are low, creeping perennial herbs with small but distinctive yellow flowers. Several species will grow in moist to very wet (but only occasionally waterlogged) soils; these include *G. elongata, G. humilis, G. lanata* and *G. ovata*. All species mentioned will grow in full sun or with moderate shading, and most can be propagated by division, especially of layered portions; for *G. ovata,* use cuttings. (Goodeniaceae)

**Gratiola** Australian Brooklime (*G. peruviana*) is a slightly succulent-looking, sprawling plant of wet places which is both drought-tolerant, and will also grow permanently in water up to 15 cm deep. The flowers are not spectacular, resembling small snapdragons. Full sun or dense shade are acceptable, and propagation is by division. (Scrophulariaceae)

**Grevillea** Several Grevilleas are reputed to be tolerant of wet soils but all are better grown in drier conditions. None are actually wetland plants in nature, although *G. rivularis* has acquired a reputation as one because of its slightly deceptive name (referring to the natural habitat in the vicinity of a river at Carrington Falls). Another species which is supposedly wet-loving but is generally healthier if kept drier is *G. shiressii*. On the other hand, *G. acanthifolia* will tolerate permanently moist to wet soils, but not waterlogging. (Proteaceae)

**Gunnera** The only native member of this interesting, circum-Antarctic genus is the Tasmanian *G. cordifolia,* a stoloniferous herb in very wet places with smallish, rounded leaves. Although not showy, it is a distinctive, densely-growing groundcover which spreads at a moderate rate in sunny locations and will tolerate drier conditions once established. Propagation is by division, although viable seed may also be set. (Haloragidaceae)

**Helichrysum** The recently discovered Swamp Everlasting (*H.* sp. aff. *acuminatum*) was a complete surprise in terms of growth habit for this genus; it is found in very wet areas which flood regularly. This plant (at the time of writing, still to be officially named) has large, yellow flowers which last just as well dried as those of any other *Helichrysum,* and the suckering stems will actually float out into water and put out roots if it is grown at the water's edge. I haven't tried growing the Swamp Everlasting in drier conditions, but it is likely that it will prove to be reasonably drought-tolerant once established. Full sun and constant moisture are needed for best growth and flowering. Plants available at the present time are being propagated by division, though it should prove easy from seed as well. (Asteraceae)

*Helichrysum* unnamed species

***Hydrilla*** Water Thyme (*H. verticillata*) is closely related to the introduced and very weedy *Elodea canadensis,* and the two aren't all that easily told apart. The simplest differentiating character is that Water Thyme usually has four or more leaves at each node, while *E. canadensis* usually has three; however, plants of both species are fairly easily recognised once you have seen living material. This is a warmth-loving species which requires fairly deep ponds in sunny or lightly shaded locations, and may not survive for long in areas where temperatures remain low for long periods of time; any piece will grow if conditions are to its liking. (Hydrocharitaceae)

***Isachne*** Swamp Millet (*I. globosa*) is a medium-height, densely-clustering grass of shallow waters and wet soils. It is not particularly distinctive except when flowering, at which time the tiny, shimmering flowers (and later seeds) held above the foliage give a light and ethereal effect. Swamp Millet will grow in full sun or with some shade, and is easily propagated by division or from seed. (Poaceae)

***Isolepis*** Formerly included in *Scirpus,* plants of this genus show a wide range of growth habits. *I. fluitans* is a fine-leaved grassy, floating plant forming dense hair-like masses in moderately deep to shallow waters, growing just as vigorously whether they are still or moving. *I. inundata* seems to grow best in shallow water, where it forms a widely spreading carpet of pale-green, near-upright stems protruding up to 10 cm above the surface, giving an effect of a uniform but carelessly cut lawn. *I. subtillissima* is a very fine-leaved species up to 10 cm high. It forms delicate-looking but tough carpets in wet and waterlogged soils, and tolerates regular flooding. All three species will grow in sun or moderate shade.

*I. nodosa* forms a more upright tussock and prefers moist to wet conditions, only tolerating brief periods of flooding (it is not abundant in sandy soil near the coast). This is another drought-tolerant Australian plant with stiffly erect, slender cylindrical stems bearing pea-sized flowerheads. Other *Isolepis* species may be available occasionally; if in doubt about their cultivation requirements, start the plants in sunny or lightly shaded sites in wet or waterlogged soil before testing them in waterlogged conditions. (Cyperaceae)

***Isotoma*** Swamp Isotome (*I. fluviatilis*) is a widespread, small-leaved prostrate herb which varies considerably across its range, and may need to be split into several species to accommodate some of the very distinctive forms known. As it will grow in moist soils as well as in shallow water, a number of unnamed forms have been cultivated for some time as ordinary garden plants. The best of these are free-flowering, with particularly well-coloured or unusually large flowers in bright- or pale-blue, lilac or, occasionally, near-white. Leaves may be very small and elliptical, elongated or completely circular and serrated. The growth habit of the plant itself may also vary from completely prostrate to semi-upright. Sun or shade are suitable and propagation is by division, though many forms come true from seed because they are selected geographical variants. (Lobeliaceae)

*Juncus* This genus is the very essence of rush; erect cylindrical stems without leaves, yet often weedy plants which spread very freely by seed. However, some smaller species are surprisingly leafy and cannot be recognised as a *Juncus* until they flower; *J. planifolius* with bright-green, red-tipped leaves is my favourite, but is a fairly short-lived plant which propagates itself by seed. *J. caespiticius* is plainer with small, chocolate-and-white flowers held above the leaves. Both these leafy varieties are found in shallow water and wet soils.

More typical species include *J. pallidus* with pale flowerheads and stiffly erect stems to over 1 m high, and *J. pauciflorus* which is similar but slender-leaved, growing to only half this height. *J. sarophus* and *J. usitacus* are closely related, also quite slender and with a slightly weeping habit, reaching the same height as *J. pallidus* but more open and delicate

*Juncus sarophus*

in appearance. All these species self-seed freely and like lots of sun. They will also grow in shallow water but are normally found in areas which only flood temporarily.

If you like the tall and upright effect of typical *Juncus* species but not their self-seeding habit, *J. ingens* is an ideal choice. This is the tallest *Juncus* in the world, reaching up to 5 m in height but more frequently around 2–3 m. Unlike other species cultivated, plants are not hermaphrodites capable of fertilising themselves; they are either male or female. By growing male plants only you can be sure that seed will not set, but females may occasionally cross-pollinate with other species under artificial conditions. *J. ingens* is a truly aquatic species unlike most other *Juncus,* and should be grown in shallow water or waterlogged soils in very sunny conditions. (Juncaceae)

*Lemna* See Duckweeds.

*Leptospermum* Tea-trees are well-known to native-plant enthusiasts, and need not be described in any detail here. A number are happy in moist to wet conditions year-round, including *L. flavescens, L. humifusum, L. juniperinum, L. lanigerum, L. liversidgei, L. myrsinoides, L. phylicoides, L. scoparium* and *L. squarrosum.* All these species will tolerate flooding for short periods, and many will be quite content even when inundated for weeks at a time. (Myrtaceae)

*Lilaeopsis* A small, untidy and rather fleshy-looking grassy plant for shallow water and wet soils, *L. polyantha* will grow in sun or very shaded locations. This is the plant for giving a natural, rather than an ornamental, appearance to your garden. Its adaptability makes it a suitable if somewhat sprawling cover among larger plants. (Apiaceae)

*Lobelia* Two Australian species of *Lobelia* are suitable groundcovers for moist to wet conditions and are not harmed by temporary flooding. *L. alata* has small lilac flowers on a low-growing, widely spreading plant, while *L. quadrangularis* is purplish-blue flowered and not as sprawling. Both grow well in full sun if kept wet and are propagated by division. Other moisture-loving species are abundant locally, but are not available through nurseries at the time of writing. (Lobeliaceae)

*Lomandra* *L. confertifolia* is a yellow-flowered, grassy tussock related to grasstrees. It is very tolerant of wet (but not waterlogged) conditions, though it is not naturally associated with water. Other species such as *L. obliqua* seem similarly versatile. All should be grown in full sun or light shade and are propagated by division, although they are relatively slow to multiply this way. (Xanthorrhoeaceae)

**Lycopus** Australian Horehound (*L. australis*) is a bold and distinctive foliage plant over 1 m high for moist to waterlogged soils, and is tolerant of prolonged flooding. Pairs of opposite, serrated leaves are closely packed along the stems, always at 90° to the previous pair like many of the fascinating, succulent *Crassula* species; the flowers are so small as to be almost invisible. Dense clumps several metres wide may form after many years of growth. The main disadvantage of Australian Horehound as a specimen plant is that it dies down in winter. Full sun or moderately shaded sites are suitable, and propagation is by division. (Lamiaceae)

*Lycopus australis*

**Lythrum** Purple Loosestrife (*L. salicaria*) is a shrubby and ornamental herb with long columns of bluish-pink flowers in summer. It spreads freely by seed and has established itself in many countries around the world to the extent that not everyone is convinced it is an Australian native. This species will grow in shallow water or any well-watered soil, and although fairly shade-loving will grow in full sun if kept wet at all times. Propagation by seed is slow, so plants are often propagated from softwood cuttings during the growing season. *L. hyssopifolia* is a much smaller, creeping species which can be cultivated in the same way; the flowers are insignificant, lilac spots along the stems. This is supposed to be an annual, but many forms I have grown are woody and long-lived. (Lythraceae)

**Ludwigia** Water Primrose (*L. peploides* ssp. *montevidensis*) forms a tangled thicket of fleshy leaves on papery, fibrous stems. It will form a low thicket on wet soils, but once planted in shallow water the floating stems trail a considerable distance out from shore. The glossy, yellow buttercup-like flowers are freely-produced in warm weather, more so in full sun than in shade. Any cutting will strike if floated in water during the growing season. (Onagraceae)

***Marsilea*** Nardoos are fascinating; the species commonly cultivated all look very much like four-leaf clover, but these are actually strange ferns, not flowering plants. Rainbow Nardoo (*M. mutica*) is variable in the wild but only one form seems to be cultivated; in this form, the leaves are beautifully patterned, shining dark-green with neat markings in shades of ochre and red. Rainbow Nardoo will grow in deep or shallow water, and will survive but not thrive on waterlogged soils.

*Marsilea* species: (left to right) *M. mutica, M. drummondii* (leaf of plant grown in wet soil), *M. drummondii* (crenate form grown in water), *M.* unnamed species (previously confused with *M. angustifolia*)

Common Nardoo (*M. drummondii*) is famed mostly for failing to provide adequate sustenance to the survivors of the Burke and Wills expedition (the spore capsules were ground to form a nutrient) and for a bland flour which could perhaps be used to advantage in some slimming diets. As a garden plant it is a most neglected native considering that it can be grown in any well-watered soil, including shaded sites. Grown in shallow water, the leaves are smooth and often autumn-leaf coloured, while in moist soils they grow larger and develop a thick coating of silver hairs. The best form in cultivation is a crenate-leaved clone I found near Echuca, Victoria. Grown in moist soil this offers all the good points of the typical plants, but in the water-grown phase the edges of the leaves are scalloped and it can be particularly colourful. This clone is also very fast-growing without becoming weedy, as it is rarely more than 20 cm high.

A small-leaved, densely-growing *Marsilea* in cultivation has previously been known as *M. angustifolia* but is actually an unnamed species (the true *M. angustifolia* is confined to northern Australia). This miniature, very clover-like fern can be grown like *M. drummondii,* but needs shallower water if kept permanently submerged. All species will grow in sun or shade and are propagated by division. (Marsileaceae)

**Mazus** Swamp Mazus (*M. pumilio*) forms low mats of crinkle-leaved rosettes in wet, often waterlogged positions. The small flowers are white or bluish. Plants are best grown in full sun and are usually propagated by division. (Scrophulariaceae)

**Melaleuca** This is another genus which is very familiar to anyone who grows Australian plants, so I will only mention that most species grow best in permanently moist to wet soils. Some, such as *M. suberosa, M. thymifolia* and *M. violacea* do particularly well in really wet conditions, while species such as *M. armillaris, M. ericifolia, M. halmatutorum, M. neglecta, M. quinquinervia, M. squamea* and *M. squarrosa* may adjust to growth in permanent shallow water. Most are grown from seed, though some species have serious weed potential. *M. quinquinervia*, in particular, is one of the most aggressive weeds of swamplands in Florida, USA. (Myrtaceae)

**Mentha** Two native mints are almost as water-loving as some of the introduced species, thriving in moist to wet soils — waterlogging is also tolerated for days or even weeks. Both *M. australis* and *M. diemenica* form low thickets of upright stems with tiny leaves and small bluish-pink flowers. They will grow in full sun or quite shady conditions, and propagation is by division. (Lamiaceae)

**Mimulus** Native Monkey-flowers are very low-growing mat-forming plants with colourful (if rather small) flowers in shades of lilac, pink or blue. *M. prostratus* and *M. repens* will grow in shallow water or on any wet soils, and are even fairly tolerant of drying-out and salt. *M. gracilis* is more showy than the creeping species as it holds the flowers up to 30 cm above the soil level, and may flower for many months. All species do best with full sun (although they may flower even in semi-shade) and are propagated from cuttings or by division. (Scrophulariaceae)

**Monochoria** Although sometimes described as a reasonably hardy plant, *M. cyanea* is sensitive to frost and will only grow well in very protected, warm sites in the warmer parts of southern Australia. This species looks rather like (and is related to) the noxious introduced weed

*Eichhornia,* yet is nowhere near as fast-spreading; the flowers are also considerably smaller and are blue or sometimes white. Shallow water or waterlogged soils in full sun or light shade are optimal, and plants are easily propagated by division. (Pontederiaceae)

*Montia* Water Purslane (*M. australasica*) is a fast-spreading, fleshy-leaved creeper for shallow water or wet soils which can be quite invasive in some conditions without established competitive plants. It will also trail into deeper waters but will not flower there, and looks uninteresting grown this way. The flowers are small and pure white but appear so thickly on plants in wet soil that the overall effect is quite showy. Water Purslane prefers full sun and will not flower well in too-shady conditions; any cutting will take root very easily. (Portulacaceae)

*Muehlenbeckia* Tangled Lignum (*M. florulenta,* previously *M. cunninghamii*) is a strange, almost leafless shrub which can form dense thickets nearly 3 m high and spread a considerable distance. It is hard to describe why it should be an attractive species yet it is, particularly when the twigs are covered in small, creamy-green flowers. Although it will tolerate prolonged and deep flooding, Tangled Lignum grows best in moist to wet soils with full sun or moderate shade, and will also survive severe droughts. Cuttings will strike, but not readily; the best means of propagation is to divide smaller, less-woody clumps. (Polygonaceae)

*Myriophyllum* Water Milfoils are a major component of wetland floras throughout Australia. Until 1985, many quite distinct species were jumbled together and shared names with unrelated South American species; however a revision was published in that year which recognises many more species, and regards all native species as quite distinct from the South American ones.

The *Myriophyllum* species mentioned in this book are all truly aquatic and grow in deep to shallow water. Many will also survive on wet soils and some are quite drought-tolerant. The much-divided, feathery underwater foliage is quite similar in many species and so they often can't be distinguished by these leaves alone. Above water, the changes in foliage are usually quite dramatic, and it is this emergent foliage and the flowers that are the most important features for identification. All species will grow in sun or semi-shade and are usually propagated by cuttings, which strike readily.

*M. propinquum* is an old name used to lump together a whole complex assortment of moderately similar species. Of these, *M. papillosum* is probably the most widely spread inland species. It is a large and robust plant often with reddish stems decorated by spirals of pink flowerlets on female plants. *M. papillosum* 'Red Dragonfly' is a very striking variety I selected over several years from particularly red clones, and have named after my nursery. For most of spring the upper side of each flattened

*Myriophyllum verrucosum*

leaflet turns a deep, crimson-red; by contrast, the underside remains bright fresh green but is completely invisible unless you bend the stems to look beneath. This is the only named selection of *Myriophyllum* in the world to my knowledge, and it thoroughly deserves this distinction because it is the most richly coloured *Myriophyllum* I know of in spring (it may change to the same colour again in autumn).

Other refugees from the *M. propinquum* label are the closely related *M. variifolium, M. crispatum* and *M. simulans*. These tend to intergrade over some of their range, but in some fairly distinct populations *M. variifolium* may tend to produce more emergent foliage whereas *M. simulans* tends to stay in the feathery, underwater phase with only small tufts of emergent foliage. The form of *M. crispatum* I have grown is distinctive for its reddish colouration. While not related to the 'propinquum' group *M. muelleri* also produces mainly underwater foliage of a much darker green than the bright, fresh foxtails of *M. simulans*; it is also much slower growing

Another group of species has a very different type of emergent leaf: rather flattened and broad, overlapping along the sprawling stems. *M. caputmedusae* (previously lumped with other species under the name *M.*

*elatinoides*) has silvery-grey, simple, scale-like emergent foliage. *M. verrucosum* is a purplish-pink version, with each leaf toothed or scalloped to varying degrees. The latter species also shows the greatest weed-potential of all native *Myriophyllum.* It can quickly choke out dams and grows into very deep water, but is easily controlled in smaller water bodies by thinning. It also seems to lose its vigour after a few years in one location and may become quite manageable even in dams after this time. Despite this warning, for unknown reasons *M. verrucosum* can also be quite difficult to establish in some waters. The species described above are the only ones in cultivation at the present time, but other selected forms and species will undoubtedly become available with time. (Haloragaceae)

*Nelumbo* The native form of the Sacred Lotus is included with non-native forms in the next chapter.

*Nertera* Bead Plant (*N. granadensis*) is a tiny-leaved, very prostrate creeper for moist to wet soils, and will tolerate flooding for short periods of time. The flowers are small but are followed by masses of orange berries which almost cover the plant completely. Bead Plant is best grown in full sun and can be propagated by division or from seed; the main problem in cultivation is that it is easily choked out or taken over by more vigorous plants. (Rubiaceae)

*Nymphaea* No Australian waterlilies occur much further south than northern NSW, and all species and forms I have tested require much warmer conditions for growth (let alone for flowering) than even a climate as warm as Sydney's can provide. Furthermore, Australian *Nymphaea* species are more difficult to transplant than the easily handled hybrids, and many are surprisingly short-lived. A number of supposedly native tropical waterlilies have been offered from time to time, but the origins of these have long been forgotten and there is no guarantee that they were originally from this country; many were obtained from overseas sources without adequate information. It is possible that selected forms can be developed from seedlings grown and tested for cold-tolerance on a very large scale, though whether such strains would attract enough interest (or be distinctive and hardy enough) to be worth the effort is questionable. (Nymphaeaceae)

*Nymphoides* The medium-sized, waterlily-like leaves of Marshworts float above the anchored runners, forming thick carpets in shallow to moderately deep water. The flowers of many species are small, yet are usually produced in quantity over a long season in warmer weather, and are always attractively tasselled. Established plants are tolerant of sun or semi-shade, and are easily propagated by breaking off the new plants which form at nodes along the runners.

Mosaic Marshwort (*N. crenata*) has pure-yellow flowers and serrated edges to the leaves which can appear quite jagged on some forms. The most interesting variations cultivated are a larger-leaved, serrated form from the Murray River which is very cold-tolerant, and 'Purple Mosaic' (incorrectly 'Variegata', although it is not strictly speaking a true variegate) with striking patterning in reddish-purple — this form probably originated in south-eastern Queensland and dies back in the winter months in cooler climates. Other forms which have been available occasionally are generally small-leaved with a rather shapeless appearance as the edges of the leaves are irregular rather than mosaic.

*N. montana* is a conspicuously fringed, yellow-flowered species previously confused with *N. geminata*. As it grows naturally at higher altitudes than other species it will survive even cooler conditions than the other native *Nymphoides* in cultivation. *N. indica* is a more subtropical species and needs relatively warm winters free of frost; its pure-white flowers are fringed with a golden-yellow centre, and the enormous leaves are often splashed with purple. (Menyanthaceae)

*Nymphoides crenata*

**Ottelia** Swamplily (*O. ovalifolia*) is the nearest thing to a southern Australian waterlily, with attractive, oval leaves tinted purple underneath. The large white flowers are three- or four-petalled (I am presently experimenting with seed-grown plants from a very attractive double form with 8–10 petals), fragile and reach up to 6 cm across with a contrasting centre of orange-red and gold. Swamplily belongs to a basically tropical genus and may not survive extreme cold, yet it has colonised most of mainland Australia (although populations in the south tend to be patchily distributed).

This plant prefers shallow to moderately deep water in open, sunlit places, and is supposed to be edible in all parts (I haven't tried it). It grows best in rich soils similar to those preferred by waterlilies and can only be propagated from fresh seed. The casualty rate is high among young seedlings which have not reached a suitable size to survive the first winter's dormancy in southern Australia, but more mature and established plants will live for many years and reach very large sizes. (Hydrocharitaceae)

**Patersonia** Purple Flags or Native Iris are small, tussocky plants sporting relatively large and showy, iris-like flowers. Not all species are water-loving, but the two most frequently available occur in heathlands which become very wet or even waterlogged in winter and spring. Both have purple flowers, with *P. occidentalis* generally a little larger and broader leaved than the slender *P. fragilis*. Full sun or light shade are suitable, and although both species will tolerate waterlogged conditions in winter they are best kept drier in the warmer months to avoid problems with fungal diseases; both species are drought-tolerant once established. Propagation is either from fresh seed or by division. (Iridaceae)

**Persicaria** Knotweeds (previously *Polygonum* species) may be creepers or upright, herbaceous shrubs that grow best in moist to waterlogged conditions, and are tolerant of prolonged and very deep flooding. *P. decipiens* is the most common species, growing as a dense groundcover or low, sprawling shrub, with leaves conspicuously marked with purple in the centre; new growth may be red or purple in some forms. *P. praetermissa* is a similarly marked creeper, yet the leaves are smaller, narrower and shaped like an arrowhead. *P. lapathifolia* is exceptional in that it is usually annual, self-seeding freely in suitable locations. Leaves of this species are rather plain but large, and so it is grown for the long and pendulous columns of creamy-pink flowers which look particularly striking on taller strains (up to 2 m high on occasion). All species will grow in full sun or semi-shade providing they are kept reasonably wet, and perennial species will grow from cuttings or divisions taken early in the growing season. (Polygonaceae)

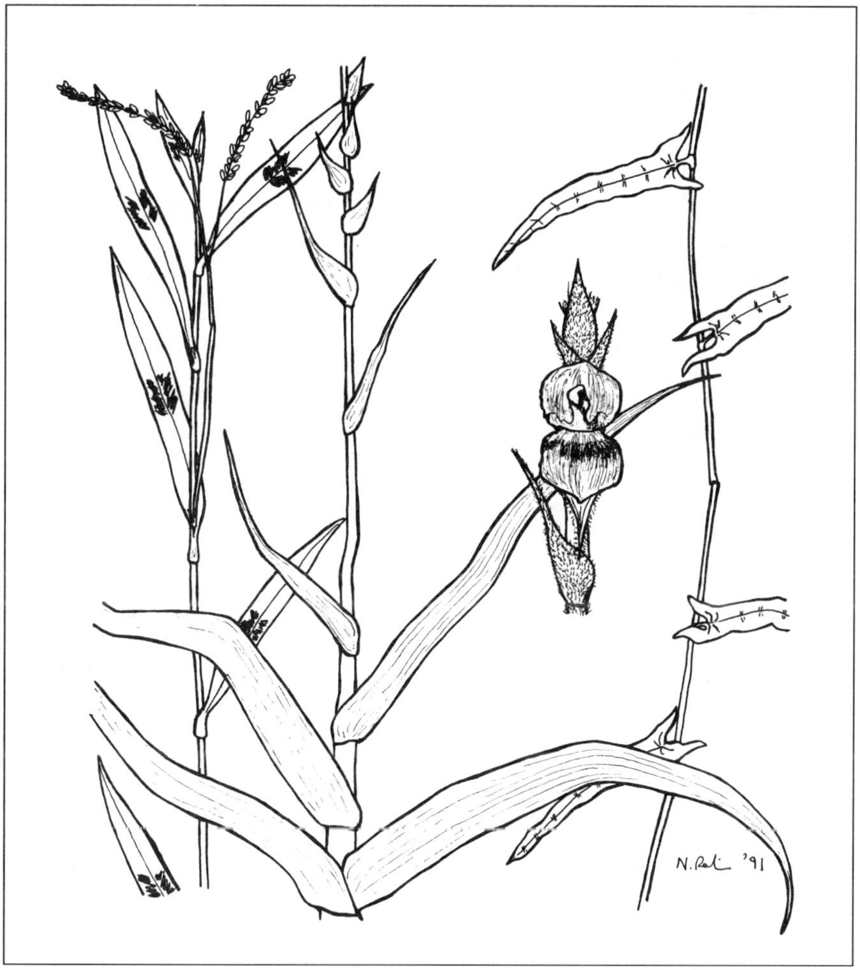

(left to right) *Persicaria decipiens*, *Philydrum lanuginosum* (base of flower stem), *Philydrum* flower (shown at twice the scale of the plant), *Persicaria strigosa* (as sometimes sold, but probably correctly *P. praetermissa*)

---

*Phaius* Swamp Orchid (*P. tankervilliae*) occurs naturally from northern NSW northwards; other species (not necessarily distinct) are named but are too rare and expensive to test for hardiness in southern Australia. Surprisingly, this red-and-white flowered orchid with blossoms to 10 cm across is tolerant of occasional, light frosts once established; however, it does best in a warm and sheltered location in most of southern Australia. Although the roots may sit in water in its natural environment, it is best kept moist to wet only over the cooler months if grown outside the natural range. Unlike most orchids, *Phaius* species may produce seedlings around the parent plant once they are well-established, but propagation is otherwise by division. Shade from the hottest summer sun is advisable wherever you grow it. (Orchidaceae)

*Persicaria lapathifolia*

***Philydrum*** Woolly Lily (*P. lanuginosum*) is a very distinctive, broad-leaved tussock basically tropical in distribution, but once also found in many cooler parts of Australia. A number of wild populations recorded in the past have apparently disappeared — due possibly to the fact that they are eaten by cattle and perhaps other livestock — but in some more northern areas they have apparently held their own. The plant itself is rather like a soft and fleshy, succulent iris, producing tall columns of small, yellow flowers on woolly flowerstems which open a few at a time over several weeks. Woolly Lily will grow and flower in full sun or quite dense shade, in any moist to wet soil or shallow water. Propagate by seed or by division. (Philydraceae)

***Phragmites*** Common Reed (*P. australis*) can be seen and grown almost anywhere in Australia. The slender, bamboo-like stems with broad, grassy leaves form dense thickets in any wet conditions from deep, brackish water to temporary wetlands where the soil may become almost bone dry in summer, and even in quite deep shade. In permanently wet, very sunny areas plants may be up to 3 m high (although this depends to a degree on the clone propagated), and topped with big and very showy, semi-weeping flowerheads as variable in colour and size as the introduced Pampas Grass (*Cortaderia*). Fast-spreading and weedy, Common Reed is nevertheless a handsome and dramatic plant which makes excellent waterfowl habitat, has edible shoots and can be used as thatch if you live in a bushfire-free area (which is to say, not in most of Australia!). Propagation is usually by division as seed may not germinate even when fresh, but divided plants may be slow to establish. (Poaceae)

*Phragmites australis*

***Potamogeton*** Pondweeds are mostly plants of deep to shallow waters, growing in sun or shade and tolerant of some degree of salinity. The underwater leaves are arranged in pairs or alternately along long and fairly brittle stems, with some producing simple, lance-shaped, floating leaves once the stems reach the surface. It is possible that some of the quite distinct plants presently lumped together as species actually encompass a much larger range of distinct taxa — hopefully, these will be sorted out in the next few years. Fully submerged species usually require deeper water and include *P. crispus,* with beautifully undulating edges to the leaves; *P. ochreatus*, with long leaves, ochre-tinted when seen out of water; and *P. perfoliatus,* the most attractive of all, with broad, tapering, elegantly marbled leaves in jade-green.

Floating-leaved types include the subtropical *P. javanicus* with tiny, bright-green, floating leaves. Although warmth-loving, this species will survive winter in climates with not too much frost. *P. tricarinatus* is more widely spread and hardy, and comes in a number of 'forms' that differ dramatically in their leaves — these are being studied and many will undoubtedly prove to be distinct species. At one extreme, a form from Kangaroo Island (SA) has floating, slightly lance-shaped leaves almost as small as those of *P. javanicus* in slowly-running water, while the very different form once known as *P. sulcatus* has very large, oval leaves up to 12 cm long. Apart from *P. perfoliatus,* all Pondweeds can be fairly invasive and very dense-growing under the right circumstances, but usually reach a maximum population in their chosen season and die away to much more moderate proportions soon after. Propagation is by division, although cuttings will take if planted early enough in the growing season. (Potamogetonaceae)

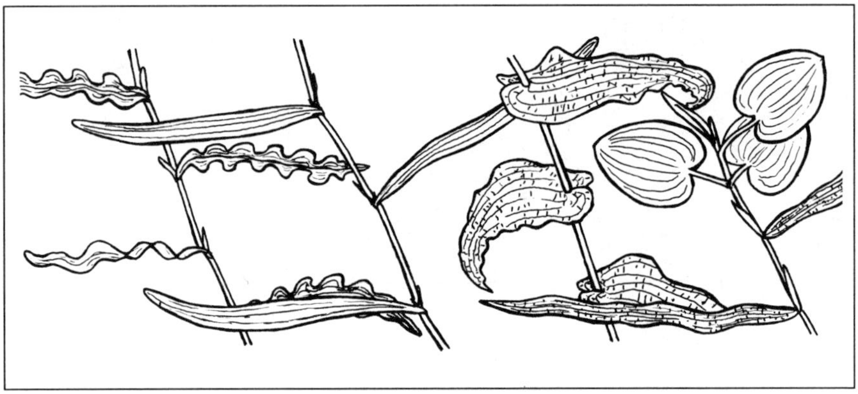

*Potamogetons*: (left to right) submerged leaves of *P. crispus*, *P. ochreatus*, *P. perfoliatus*; and *P. tricarinatus* (right) showing submerged leaves and three floating leaves.

*Pratia* These are creepers with tiny serrated leaves and small, palmate flowers. Although attractive enough they may spread very quickly and become a weed which is impossible to eradicate. Two species (*P. pedunculata* and *P. purpurascens*) seem to prefer growing in moist to wet soils, and are easily propagated by division.

*Ranunculus* Most native buttercups are associated with wet places to some degree, but several species or species groups are aquatic. The taxonomy of these is currently in a somewhat confused state, with no improvement in sight in the near future, and so I have not attempted to separate the many similar-looking forms which are distinguished primarily by their flowers. All will grow in sun or semi-shade and should be propagated by division because of the strong possibility of hybridisation; indeed, the complex droves of closely related forms may be partly natural hybrids.

River Buttercups are the most commonly seen species, and the forms in cultivation are probably either *R. inundatus* or *R. amphitricus* (previously *R. rivularis*). The leaves of any species in the *R. rivularis/R. inundatus* group are divided to varying degrees, from being relatively coarsely lobed to finely and intricately dissected ones which give the impression of lace filigree work. These plants are grown mainly for their

*Ranunculus inundatus*

Leaves of several native *Ranunculus* species, showing some (but by no means all) of the variations which can be found. Variations in leaf-type overlap considerably, so it is rarely possible to identify *Ranunculus* by leaves alone. (left to right) *R. amphitricus*, two forms of *R. inundatus*, and *R. trichophyllus*.

foliage variations as the flowers are often small or even inconspicuous. Submerged leaves are often larger and more finely divided than those seen above water because they don't have to support their own weight, and so can elaborate and divide as they see fit. As plants grown for too long in deeper water often lose vigour, it is best to either use these buttercups where water levels drop naturally, or plant them permanently in shallow water or wet soil.

Water Buttercup (*R. trichophyllus*) is exceptional in that it is a bona fide aquatic which doesn't like even shallow water for any length of time. The leaves are so soft and finely divided that they tend to fold together, giving a general impression similar to the introduced (potentially weedy) *Cabomba,* for which this species makes a salt-tolerant substitute. It is worth noting that some authorities now regard this as an introduced species.

There is also a group of native alpine *Ranunculus* originating from very wet places (e.g. *R. anemoneus*), but these are snowmelt plants which are difficult or even impossible to keep at low altitudes in this country, although some may survive for a few years in the much colder conditions of northern Europe when grown from seed. (Ranunculaceae)

*Restio* Australia boasts a number of interesting and attractive members of the Restionaceae, but the only ones commonly available in nurseries belong to this genus. *Restios* often have the appearance of very slender, smallish bamboos, with distinctly jointed segments from which the foliage springs; but unlike many bamboos they will not run from where they are planted, forming neat tussocks instead. The best-known species is *R. tetraphyllus,* a widely spread plant occuring in two distinctive and attractive subspecies: *R. tetraphyllus tetraphyllus* may reach over 3 m high when very old, but is usually an upright tussock half that height with soft, feathery, semi-weeping foliage towards the top; *R. tetraphyllus meiostachyus* is a shorter, more feathery and open tussock reaching up to 2 m in height, but the clone usually available at nurseries only reaches around 1 m when not flowering (the flowerstalks are taller).

The other species in cultivation are smaller, more compact and less striking, yet have their own interest if you like grassy-looking plants generally. *R. australis* is a particularly deep-green and is probably the most cold-tolerant of the genus for it is sub-alpine in parts of its range. Both *R. complanatus* and *R. tremulus* have distinctly flattened stems; the latter is also rather twisted and convoluted in some clones grown. All species will grow in full sun or shade and prefer moist to wet soils though they will also grow in shallow water. Propagation is by division or from fresh seed; male and female plants are needed if any is to be set. (Restionaceae)

*Riccia* Lace liverworts are fine, intricately branched floating plants which do best in shaded areas or amongst other vegetation. They will also grow on very wet soils in shade, forming dense and mossy-looking, bright-green tufts. There are probably several very similar-looking species in Australia, but only the very uncommon Victorian form is currently in cultivation; this is fairly slow-spreading unlike at least one (probably non-native) species in the aquarium trade. (Ricciaceae)

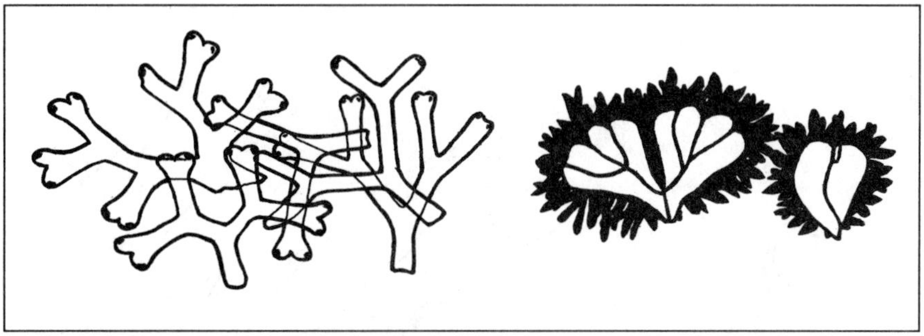

*Riccia* species (left) and *Ricciocarpus natans* (right), both magnified.

*Ricciocarpus* Bearded Liverwort (*R. natans*) is another unusual floating plant, forming pairs of bright-green 'leaves' from which hang tufts of matt-black rootlets which give the effect of a dense beard. Unlike the *Azolla* species, Bearded Liverwort is not particularly fast-spreading, and should be protected from the hottest sun. (Ricciaceae)

*Rumex* While Water Dock (*R. bidens*) is obviously a dock, the upright, bright green, spearhead-shaped leaves are much neater than those of most non-native species. In rich soil, Water Dock will form dense and rather handsome clumps with the hollow, floating stems trailing short distances out over deeper water. Plant smaller pieces in shallow water or wet soil and allow them to spread from there, whether in full sun or semi-shade. Propagate by division. (Polygonaceae)

*Samolus* Creeping Brookweed (*S. repens*) is a small creeper for very shallow water and wet soils. It is able to tolerate full sun or quite dense shade, and high salinities. The leaves are rounded and very dark green; in warmer weather neatly symmetrical, pure-white flowers top the rather sprawling stems. Propagate by division. (Primulaceae)

*Schoenoplectus* This is another genus which was previously a part of *Scirpus,* and three of its species are presently in cultivation. *S. validus* is the common River Clubrush with tall, cylindrical stems topped with small pom-poms of chocolate-brown flowerheads, and starchy, edible rhizomes. This is a moderately fast-spreading plant for full sun or semi-shaded conditions and will grow in quite deep water or any wet soil. Like the much shorter Delta Clubrush (*S. pungens*), it is a waxy blue-green colour.

Delta Clubrush is so-called because it is frequently found in estuarine conditions including river deltas, and also because of the pronounced, triangular cross-section of the erect stems. (It is the only pun I have ever inflicted in renaming a plant for commercial purposes — the old common

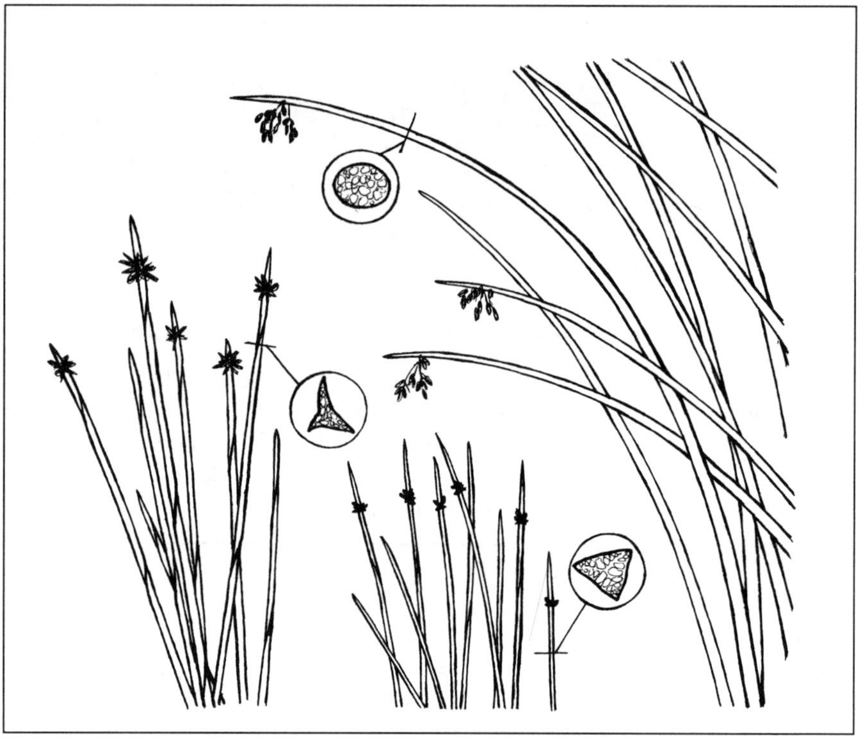

Native *Schoenoplectus* species, showing cross-sections of their stems: (left to right) *S. mucronatus*, *S. pungens*, *S. validus*.

name of American Clubrush refers to an obsolete synonym.) Cultivation is similar to *S. validus* though *S. pungens* is very salt- and alkali-tolerant, faster spreading and prefers more sun. With triangular stems, *S. mucronatus* is similar to *S. pungens* in appearance but is generally taller, more robust and a rich, dark green. All species are propagated by division. (Cyperaceae)

**Schoenus** *S. maschalinus* spreads as a dense mat of tiny, grassy leaves; the form cultivated is low-growing in full sun or moderate shade, and prefers wet or moist soils although it will tolerate both drought and flood. Propagation is by division. (Cyperaceae)

**Scirpus** This genus has been revised and all species presently in cultivation are described under *Bolboschoenus*, *Isolepis* and *Schoenoplectus*.

**Scleranthus** *S. biflorus* forms a moss-like cushion and, as the flowers are greenish and very small, I suppose we can stretch the truth and call it a foliage plant. Full sun and moist to very wet (occasionally waterlogged) conditions are ideal, and plants are propagated from cuttings or by division. (Caryophyllaceae)

**Selaginella** Swamp Clubmoss (*S. uliginosa*) looks like a tallish moss, forming dense patches of pale, bright-green foliage in sunny and lightly shaded sites. Although it prefers moist to wet sites it will not survive prolonged flooding. Propagation is by division, but should only be practised on well-established plants so that all pieces separated have a reasonable root system; it is slow to establish even with the greatest care. (Selaginellaceae)

**Sparganium** Floating Burr-reed (*S. antipodum*) is a fleshy-leaved reed with interesting, rather zig-zag flowerstems carrying rounded, burr-like flowerheads. Plants do best in moderately deep water, but are not greatly affected by further flooding and will also grow on wet soils. Full sun or semi-shade are suitable, and propagation is by division unless you can obtain fairly fresh seed; this species is moderately cold-sensitive and is probably best grown in coastal and warmer areas. (Sparganiaceae)

**Spirodela** See Duckweeds.

**Sticherus** Fan Ferns are very closely related to *Gleichenia* and are cultivated in much the same way although they may tolerate shade a little more than the Coral Ferns. The general appearance of their branching is also the same, but the fronds of Fan Ferns are a much paler, brighter green and are not divided to the same degree, so they don't give the impression of being made up of chains of tiny beads. The three species most likely to be available are *S. flabellatus, S. lobatus* and most frequently *S. tener*. Propagation from spores is reasonably easy, but slow. (Gleicheniaceae)

**Todea** King Fern (*T. barbara*) is a very large-growing, low-trunked species sometimes confused with the true treeferns. Although the trunks rarely exceed 1 m in height they can be very wide; in the natural swamp on our property some old King Ferns are so large that I can only stretch my arms halfway around them, but these may be the result of two or more crowns grown together over centuries. This magnificent fern is easily grown in any reasonable soil kept moist to wet, and will tolerate waterlogging for long periods of time — moderate to heavy shading is best. Propagation from fresh spores is slow but not difficult; however, plants sold have almost always been stripped from the wild with much the same tender care as mentioned under *Dicksonia*. (Osmundaceae)

**Triglochin** Streaked Arrowgrass (*T. striata*) is a wiry, tuft-forming, grassy plant which spreads quickly in wet soils or shallow water. It is small and insignificant at most times, but produces carpets of neat, symmetrical, upright, beaded flowerstems. Other, related species show variations in leaf shape and size, yet in nurseries have been lumped together with *T. striata*. The larger *Triglochin* species have similarly attractive flowerstems on a much larger scale, with flowers followed by elegant, swirled fruits like tiny, green meringues.

The only larger species presently available through nurseries is Water-ribbon (*T. procera*), a glossy, broad-leaved plant with two distinct forms: semi-upright Water-ribbon is generally the broadest leaved, standing up to around half-a-metre high and drooping somewhat towards the tips of the leaves; the floating-leaf form trails long, slender leaves along the water's surface whether grown in still or flowing waters, but may become slightly upright in very shallow waters and on mud as water levels fall. Both forms have edible tubers which require vigorous digging to collect in quantity.

A related species from cold streams in East Gippsland has spiral leaves, but how well it will retain these in cultivation remains to be seen. All *Triglochin* species are best grown in full sun or light shade, are fairly drought-tolerant and are generally propagated by division. Transplanted pieces may be slow to re-establish themselves, so seed may be a viable alternative for large plantings if you have the patience to raise them this way in pots. (Juncaginaceae)

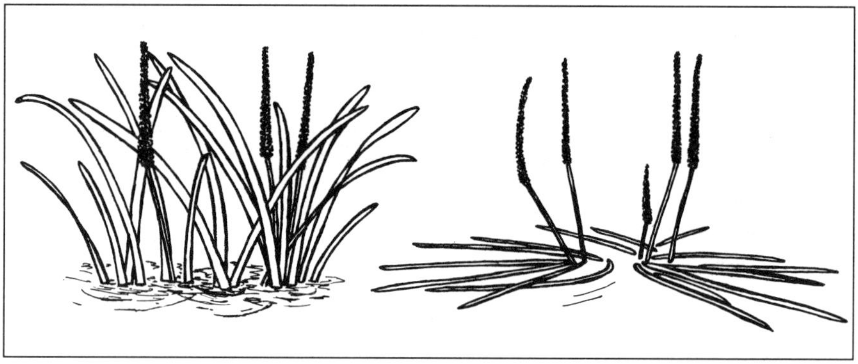

Two forms of *Triglochin procera* with distinctly different habits. The leaves of the erect form (left) may float except in the deepest water, and the floating-leaf form (right) seen growing in very shallow, still water. Leaves of this form seldom stand up even if the water dries out completely. There are also intermediate forms between these two extremes.

***Typha*** There are two species of Cumbungi (also called Bullrushes, and the origins and correct use of this name are discussed under non-native *Typha* species) native to Australia. Both species are tall-growing, fast-spreading and invasive rushes which provide excellent waterfowl habitat if you don't care whether you can see the water or the waterbirds through them. The poker-heads are ornamental if picked early enough and dried — too late and they will disintegrate into clinging fluff that sticks to every fabric it can find. The rhizomes produce a good quality, edible starch which takes a fair amount of work to process, and new shoots are edible when peeled. As they are very productive plants much use could be made of the dense stands which clog inland irrigation areas, but in this food-rich country it is unlikely that anyone will bother trying to harvest them on a commercial scale.

*T. orientalis* is generally larger and broader in every way than the smaller and more delicate *T. domingensis*; however, there is a fair amount of overlap between the two and only floral characters will definitely separate most forms. Selected forms with particularly large and bold, or slender and elegant bullrushes (according to species) are propagated by division. Both species are happy in sun or shade, in anything from deep water to drying mud, and are quite tolerant of drought and some salinity (a nice way of saying they are very weedy and will grow anywhere). Pot culture is the best way to control them, although potted plants will still spread by seed if kept in a dam or other large waterbody if seedlings aren't pulled out regularly. (Typhaceae)

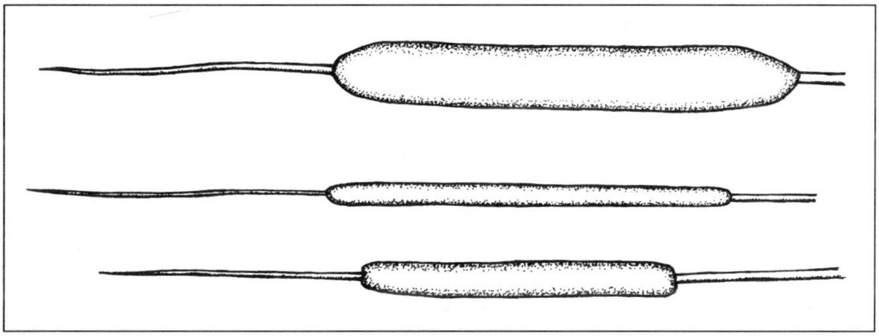

Native *Typha* species, showing the difference between a particularly large form of *T. orientalis* (top) and two slender forms of *T. domingensis* (below). There are many intermediate forms which can only be identified by details of their microscopic structure.

**Urtica** Native Stinging Nettle (*U. incisa*) is closely related to and just as attractive as the introduced nettles, yet mature plants have narrower and more elongated leaves. They are ornamental to look at but are just as fierce when it comes to biting as non-natives, and should be grown away from paths or planted where they will discourage unwanted foot traffic. Moist to moderately wet soils and sunny or moderately shaded positions are best, and flooding for a week or two at a time will not harm them. Young shoots are edible and rich in nutrients, but must be steamed or boiled for a short while as heat totally nullifies the stinging compounds. Propagate by division (but wear gloves). (Urticaceae)

**Utricularia** Bladderworts are often associated with wetlands but only a few are truly aquatic. The small but showy flowers in yellow or purple are held above water, looking as if they wear tiny aprons, but the main interest in these little carnivores is the way they trap their prey. The finely and intricately divided foliage contains many tiny bladders, actually tiny traps for minute underwater animals, and these are virtually spring-loaded to snap shut on being disturbed, drawing water and prey inside. All the aquatic species grow best floating among other plants whether in sun or shade and propagate themselves without help. The most ornamental foliage is that of the elegantly whorled, winter-dormant *U. australis*; other species are more tangled and amorphous-looking but are much more likely to flower, especially *U. gibba* which will produce dense and showy carpets of yellow flowers over a long period of time. (Lentibulariaceae)

**Vallisneria** At present there are two very variable, native Eel-grass species, but the taxonomy of this group is not well-defined and two or three more may be added at a later date. They are separated by floral and fruit characteristics but, as many forms don't seem to produce fully developed fruits, it may be difficult to identify them definitively. From a horticultural point-of-view, all forms are grassy plants of shallow to deep permanent waters, forming thickets reaching to the surface. All will tolerate full sun if planted deeply enough, or even full shade in shallower water. The various forms are sold according to how long and broad the leaves become under ideal conditions, and are propagated by division as they spread quickly by runners. (Hydrocharitaceae)

**Villarsia** Marshflowers are closely related to *Nymphoides,* and from the gardener's point-of-view are the shallow-water and temporary-pond version of that genus. All are reasonably drought-tolerant once established, and produce their yellow, usually somewhat tasselled flowers in spring as the shallows they inhabit dry out; however, permanent water will not worry them. The leaves can become very tatty in summer unless there is a reasonable amount of shade. *V. umbricola* is a small-flowered species, forming low mounds of bright-green leaves as long as it is submerged. Some forms of *V. exaltata* can have an appearance not unlike violets until the flowers appear. *V. reniformis* is the most commonly available and variable species, often with particularly large and showy flowers. Some small-leaved but interestingly shaped species such as the short-lived *V. calthifolia* and *V. parnassifolia* have occasionally been available; these will tolerate moderate frosts and will set seed freely once established. (Menyanthaceae)

**Viola** Two native violets are happy in really wet soils and are often found in locations which run with water to some degree in winter and spring. Both are best grown with reasonable shading from the summer sun, and prefer moist to wet conditions during the warmer months. *V. betonicifolia* has heart-shaped, dark-green leaves which may have a purple tint, and flowers which vary from white through various shades of blue to purple. This species spreads freely by seed to the extent that it is regarded as a weed in some nurseries. *V. hederacea* is an altogether more restrained plant and forms low carpets of small, kidney-shaped leaves through which the lilac-blue and white flowers show. Selected forms of both species must be propagated by division, although some selected wild forms will come true from seed if there are no potential cross-pollinators about. (Violaceae)

*Wolffia* See Duckweeds.

*Xyris* Yellow-eyes are small and slender-leaved, grassy, tussock-forming plants which grow in any moist to wet soil. They look like many smaller rushes and grow in similar habitats, but when the small, yellow or cream flowers appear from among the scales of the dark flowerheads there is no mistaking them. Full sun is best for growth and flowering, and propagation is from fresh seed or by division; however, divided plants will often waste away unless conditions are to their liking. The species most likely to be available are *X. gracilis* and *X. operculata* as most others are more cold-sensitive. (Xyridaceae)

# Chapter Four

# Non-native Waterplants

## Introduction

Although Australia boasts a diverse range of waterplants, their cultivation and selection is really only just beginning. Waterplants from many other parts of the world have been grown and selected by different cultures (for 1200 years in the case of *Iris laevigata*); and, as you would expect, the species chosen and developed have been the most eye-catching. There are hundreds of unusual and beautiful waterplants from outside Australia, and most of them have been imported at one time or another.

The majority of non-native varieties are no more invasive than exotic garden plants generally, but be warned: waterplants include some of the most noxious weeds in the world, and without the natural controls on their reproduction which are found in their homelands a few can spread so quickly that you can actually watch them grow! *Eichhornia, Salvinia* and non-native *Alternanthera* are the three most disastrous invaders to date, nevertheless Australia is still free of other potentially serious weeds such as *Trapa.*

No other weedy species is anywhere near as seriously invasive as those mentioned above, and I have tried to balance their weed potential against ornamental value. For example, many *Sagittaria* species are fast-spreading, and some are thoroughly naturalised in Australia (although they haven't shown any tendency to choke out native species or invade undisturbed habitats), they are weedy only in the sense that they can quickly take advantage of disturbed areas, but their ornamental value is high so many remain worthwhile plants to grow if reasonable care is taken to prevent their further spread.

There is a third group of non-native, potential weeds which are not really a threat to any undisturbed environments, and are so uninteresting that there is no reason to bother growing them. These species have been mentioned for the sake of completeness because they are still occasionally offered for sale.

The alphabetical listings of non-native plants described in this section are arranged in the same way as the Australian plants and use the same abbreviations. Synonyms are only mentioned if name-changes are either recent or are not generally known. Some plants listed are perfectly familiar to gardeners in general, so only brief mention is made of them along with a description of their place in a watergarden. The lists of plants which follow are intended to be used in the same way as those in the section of Australian plants, and the reader should refer to the beginning of that section for further information. Keep in mind that the differences between moist, wet and waterlogged soils are important if you don't want to lose your plants; these are discussed in Chapter 2.

# *Non-native plants suited to various uses and conditions*

**Edible non-native plants** can be found in the following genera:

*Acorus, Althaea, Aponogeton, Bambusa, Butomus, Camassia, Canna, Colocasia, Cyperus, Cyrtosperma, Eleocharis, Hemerocallis, Ipomoea, Nelumbo, Nymphaea, Orontium, Oryza, Peltandra, Polygonum, Pontederia, Rorippa, Sagittaria, Sium, Trapa, Typha, Typhonodorum, Veronica, Zizania.*

**Shade-tolerant non-native plants** can be found in the following genera:

*Aponogeton, Arum, Berula, Cabomba, Callitriche, Caltha, Chimonbambusa, Didiplis, Echinodorus, Egeria, Eleocharis, Fritillaria, Gentiana, Gymnocoronis, Hosta, Houttuynia, Leucojum, Ligularia, Lilaeopsis, Ludwigia, Lythrum, Marchantia, Myosotis, Myriophyllum, Nuphar, Onoclea, Ophiopogon, Phalaris, Primula, Rorippa, Rotala, Samolus, Saururus, Scirpus, Shibataea, Vesicularia, Viola, Zantedeschia.*

**Non-native plants suitable for deeper water** can be found in the following genera:

*Aponogeton, Cabomba, Didiplis, Egeria, Hydrocleis, Iris, Lilaeopsis, Ludwigia, Myriophyllum, Nelumbo, Nuphar, Nymphaea, Nymphoides, Orontium, Samolus, Thalia, Typha, Vesicularia.*

**Non-native plants suitable for shallow water** can be found in the following genera:

*Acorus, Aponogeton, Berula, Butomus, Cabomba, Calla, Callitriche, Caltha, Camassia, Canna, Cardamine, Carex, Colocasia, Cotula, Crinum, Cyperus, Cyrtosperma, Didiplis, Echinodorus, Egeria, Eleocharis, Eriophorum, Euphorbia, Heteranthera, Hibiscus, Houttuynia, Hydrocleis, Hypericum, Ipomoea, Iris, Juncus, Lilaeopsis, Lobelia, Ludwigia, Lythrum, Mentha, Menyanthes, Myosotis, Myriophyllum, Nelumbo, Nuphar, Nymphaea, Nymphoides, Onoclea, Ophiopogon, Oronyium, Oryza, Peltandra, Phalaris, Pontederia, Preslia, Ranunculus, Rorippa, Rotala, Sagittaria, Samolus, Saururus, Scirpus, Thalia, Typha, Typhonodorum, Veronica, Vesicularia, Zantedeschia, Zizania.*

**Non-native plants suitable for waterlogged soils** can be found in the following genera:

*Acorus, Arundo, Berula, Butomus, Calla, Callitriche, Caltha, Camassia, Canna, Cardamine, Carex, Chimonobambusa, Colocasia, Cotula, Crinum, Cypella, Cyperus, Cyrtosperma, Dionaea, Echinodorus, Eleocharis, Eriophorum, Euphorbia, Gladiolus, Gunnera, Hemerocallis, Heteranthera, Hibiscus, Houttuynia, Hymenocallis, Hypericum, Ipomoea, Iris, Juncus, Leucojum, Lilaeopsis, Lobelia, Ludwigia, Lysichiton, Lythrum, Marchantia, Mentha, Menyanthes, Mimulus, Moraea, Myosotis, Onixotis, Onoclea, Ophiopogon, Oryza, Osmunda, Peltandra, Phalaris, Polygonum, Pontederia, Preslia, Ranunculus, Rorippa, Rotala, Sagittaria, Sarracenia, Saururus, Schizostlis, Scirpus, Sisyrinchium, Sphagnum, Symplocarpus, Thalia, Typha, Typhonodorum, Veronica, Wachendorfia, Zantedeschia, Zizania.*

**Non-native plants suitable for wet soils** can be found in the following genera:

*Aconitum, Acorus, Althaea, Anemone, Aristea, Arum, Aruncus, Arundo, Astilbe, Bambusa, Berula, Callitriche, Caltha, Camassia, Canna, Carex, Chimonobambusa, Chondropetalum, Colocasia, Cotula, Crinum, Cypella, Cyperus, Dietes, Dionaea, Elegia, Eleocharis, Equisetum, Eupatorium, Euphorbia, Filipendula, Fritillaria, Geum, Gladiolus, Gunnera, Hemerocallis, Hibiscus, Hosta, Houttuynia, Hymenocallis, Ipomoea, Iris, Juncus, Kniphofia, Leucojum, Ligularia, Lilium, Lobelia, Lysichiton, Lysimachia, Lythrum, Marchantia, Mentha, Mimulus, Miscanthus, Mollinia, Moraea, Myosotis, Narcissus, Onixotis, Onoclea, Ophiopogon, Osmunda, Parochetus, Peltiphyllum, Phalaris, Polygonum, Primula, Ranunculus, Rorippa, Salvia, Sarracenia, Sasa, Schizostylis, Shibataea, Sisyrinchium, Sium, Sphagnum, Symplocarpus, Trollius, Typha, Typhonodorum, Veronica, Viola, Wachendorfia, Zantedeschia.*

**Non-native plants suitable for moist soils** can be found in the following genera:

*Aconitum, Acorus, Althaea, Anemone, Aristea, Arum, Aruncus, Arundo, Astilbe, Bambusa, Caltha, Camassia, Canna, Carex, Chimonobambusa, Chondropetalum, Crinum, Cypella, Cyperus, Dietes, Dionaea, Elegia, Equisetum, Eupatorium, Filipendula, Fritillaria, Geum, Gladiolus, Gunnera, Hemerocallis, Hosta, Hymenocallis, Iris, Kniphofia, Leucojum, Ligularia, Lilium, Lobelia, Lysimachia, Lythrum, Marchantia, Mentha, Mimulus, Miscanthus, Mollinia, Moraea, Narcissus, Onixotis, Onoclea, Ophiopogon, Osmunda, Parochetus, Peltiphyllum, Phalaris, Polygonum, Primula, Ranunculus, Salvia, Sasa, Schizostylis, Shibataea, Sium, Sphagnum, Trollius, Viola, Wachendorfia, Zantedeschia.*

**Non-native plants which float** can be found in the following genera:

*Eichhornia, Pistia, Salvinia.*

# Alphabetical list of non-native waterplants and wetland plants

**Aconitum** Several species of Monkshoods are available from time to time in Australia, producing columns of strangely shaped, deep-blue flowers in summer. They are usually grown as ordinary garden plants, yet many species look and grow much better in quite wet soils with shelter from the hottest sun. Established plants should be divided with care as they are very poisonous. The only water-loving species readily available is sold as *A. napellus*, but is apparently quite distinct from the true species. (Ranunculaceae)

**Acorus** There are probably two species of these rush-like plants. Both forms of the smaller *A. gramineus* will do well in shallow water, or any moist soils. 'Pusillus' is a very dark-green, fan-like plant growing to about 10 cm, while the form variously known as 'Decorata' or 'Variegata' is taller (to 45 cm) and is marked with broad, creamy stripes. A yellow-variegated form of this species called 'Ogon' has recently been imported and should be available within several years.

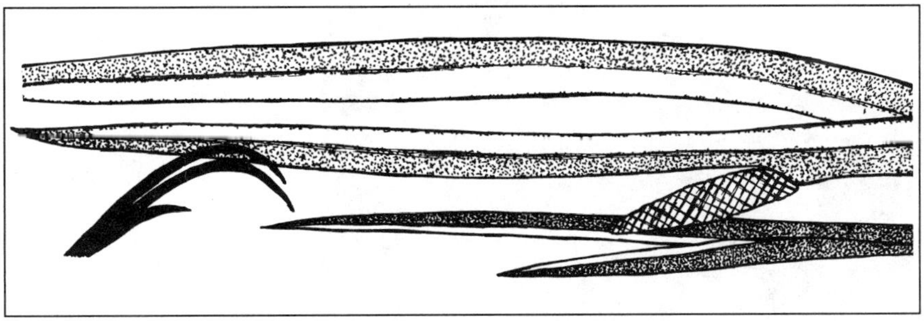

*Acorus calamus* 'variegatus' (top), *A. gramineus* 'Pusillus' (below left), *A. gramineus* 'Decoratus' (below right)

*A. calamus* is taller still with broader leaves resembling an iris, and must be grown in shallow water if it is to thrive. This is the Sweet Flag which was strewn on flagstones in mediaeval England to sweeten the atmosphere indoors, as the leaves have a strong fragrance reminiscent of spiced mandarin skins. The rhizomes have been candied as a delicacy, and various parts of the plant have medicinal uses. The wild form of this plant is grown for its useful properties, fragrance and historical interest, but there is also a variegated form with bold, creamy-yellow striping which is less fragrant. This variegated form of *A. calamus* may die down in winter but the other forms are usually evergreen. All forms are propagated by division and look best in full sun. (Araceae, but probably to be separated into a family of their own shortly.)

**Althaea** Marshmallow roots (*A. officinalis*) were the original source of the well-known flavouring of the same name, though this is now often made with synthetic flavourings. Although it has columns of attractive pink flowers, the plant is a rather non-descript member of the hibiscus family, and will grow in well-watered garden soil or quite wet (but not waterlogged) conditions; some shade is needed in hotter areas. Propagation is by seed or division, in both cases in mid-spring. (Malvaceae)

**Anemone** Two species of this well-known genus are riverside plants. *A. virginiana* grows in moist or wet soils and is reputed to tolerate flooding for short periods. The flowers of this species are very small and of a remarkably uninspiring green nowhere near as attractive as the somewhat larger, delicate pink or white flowers of *A. rivularis*. Both will grow in sun or semi-shade, and are most easily propagated by seed. (Ranunculaceae)

**Aponogeton** Water Hawthorn (*A. distachyon*) grows mostly in cooler weather, in depths from 60 cm to almost no water at all. It is a

*Aponogeton distachyon*

drought-adapted plant which dies back to a tuber and will survive complete drying out in summer. As it spreads rapidly by seed, it should not be planted near natural waterways. The flowers are curious and very attractive double-spikes of white with a delicious fragrance reminiscent of Hawthorn (hence the name), and have been used as a flavouring for mutton stew in South Africa; tubers are also edible. Unlike most other waterplants it flowers from mid-autumn through to spring even in fairly shady locations, so it is a valuable complement to the sun-loving, summer-flowering waterlilies. Propagation is usually by seed, but very old tubers may be divided in early autumn as they start to grow if they have two or more distinct crowns. (Aponogetonaceae)

*Aristea* Only one species of these sun-loving plants (*A. ecklonii*) will grow in waterlogged conditions, but several others may do quite well in fairly wet soils. These small, iris-like plants produce columns of vivid, bright-blue flowers over most of the warmer months and need full sun for best results. Propagation is by division, but seed will produce flowering-size plants quite quickly when available. (Iridaceae)

*Aristea ecklonii*

**Arum** Marbled Arum (*A. italicum*) looks best grown in shade in wet soils which are never waterlogged for too long. The form 'Pictum' mostly seen in Australia has dark-green leaves overlaid with white marbling, and green, slightly malodorous arum flowers which are followed by columns of bright-red berries. While these look very attractive, if you don't want new plants popping up everywhere cut them off before they start to fall. Although this is not a particularly invasive plant, in gardens where it is really happy it is virtually impossible to get rid of as it grows deep into the soil. (Araceae)

**Aruncus** Goatsbeard (*A. dioicus*) is a tall herb for the water's edge, but the crown must be planted well above the level of waterlogging in the soil. The foliage is attractive in a ferny way, and in spring, feathery masses of white flowers tower above it giving an effect similar to giant *Astilbe* blooms. Older clumps can be quite spectacular when flowering but look best if given some protection from the hottest sun. Goatsbeard is also sold under an old name, *A. sylvesteri*. Several smaller species are occasionally available but lack the grandeur of the larger species. (Rosaceae)

**Arundo** Bamboo grass (*A. donax*) is a handsome, bamboo-like grass growing to 4 m or more high with long, pendulous leaves and gracefully-curved stems. It grows well in full sun or with some shading but is most lush when kept thoroughly wet; it will even tolerate waterlogging. The sheer size of this widely spreading plant makes it unsuitable for smaller gardens, though there is a form with boldly white-variegated leaves which is smaller and less vigorous. (Poaceae)

**Astilbe** Usually treated as ordinary garden plants, *Astilbe* species and hybrids need constant water for best health and even tolerate water-logged soils for short periods of time. Full sun gives the best show of the plume-like flowers, which are available in a range of colours from white through to pinks, lilacs and reds. At least 20 varieties have been offered in Australia, and new varieties are constantly being released so I may not have seen the full range of colour available, but my favourite will probably always be the flamboyant red of 'Fanal'. Some of the named varieties which have been sold are so different from the original descriptions of these varieties that it is impossible to believe they refer to the same plant, so select your varieties by the descriptions on catalogues and not from European books. They look best planted in groups of one colour, but there is no reason why the groups shouldn't rub shoulders to give a colourful display in late summer. (Saxifragaceae)

**Bambusa** Most of this genus are fairly tropical bamboos, and only *B. multiplex* is both relatively small and cold-hardy. It comes in several very different forms, all of which are clumping bamboos (they won't run all over your yard!). Most forms should have as much sun as possible for good colour and form.

'Alphonse Karr' grows to 4 m in cooler climates, and easily twice that in more northerly areas. This is one of the finest of ornamental bamboos for smaller gardens; the golden-yellow stems are striped neatly but variably with a richly-contrasting green along their length. By contrast, 'Stripestem Fernleaf' has slender, similarly-coloured stems rarely more than 1 m in height, and tiny leaves. 'Rivierorum' also has small leaves, but may reach 4 m high. In cooler areas (especially in shade) it is rarely above 2 m, and the fine leaves give it a ferny appearance. As this form is unusually shade-tolerant it is one of the few bamboos which can be grown indoors, although it may produce much larger leaves in low-light conditions.

'Silverstripe' is the largest of the *B. multiplex* forms, reaching up to 13 m in warmer areas; the stems and leaves are variably striped with white. All of these forms must be propagated by division and none will tolerate waterlogged conditions for long, so they should be planted near the water's edge, not on it. New shoots of most forms are edible, but may need repeated soaking or cooking to remove bitterness. (Poaceae)

**Berula** Water Celery (*B. erecta*) is a shade-loving plant which both looks like and is related to celery, with a pleasant herby aroma. It dislikes heat and is not suited to hot climates. Dense tangles of this plant in shallow water give the appearance of a fernbrake but it can become very weedy under these conditions. Propagate by division. (Apiaceae)

**Butomus** Flowering Rush (*B. umbellatus*) has unusual, three-sided stems and looks remarkably like a rush of some kind. In summer, it gives itself away as a very different plant when it produces flowerstems above the foliage with a mop-head of pink flowers. The clone in Australia may not flower except in cooler or cold areas, and even then mostly in cool summers. Plants should be 5–10 cm underwater but they will also grow (though not necessarily flower) in waterlogged soils; full sun gives best results. Propagate by breaking off the edible bulbils on the underground rhizomes, or by division of the rhizomes. (Butomaceae)

**Cabomba** Fanwort (*C. carolineana*) is an elegant submerged plant which must have soft, preferably acid water if it is to thrive. Surprisingly, it is also a very vigorous grower if it is happy and may be a potential weed in some conditions. The fragile stems are surrounded by soft fans of finely divided leaves and pieces which are broken off take root readily. *(Cabombaceae)*

**Calla** Water Arum (*C. palustris*) is an attractive and unusual scrambling plant for shallow water and mud. It is useful for hiding the division between water and bog but dies down in winter, leaving a bare area. The leaves and flowers are similar to those of the common Arum lily (*Zantedeschia*) but much smaller. Although it needs a reasonable amount of sunlight, it should be sheltered from the hottest sun. Propagate by cuttings, or very fresh seed if you can get it. (Araceae)

*Callitriche* There are two common non-native species of Starworts in Australia, and between them they may have pushed the less invasive native species out of their niche. Both *C. stagnalis* and *C. hamulata* are plants of fairly shallow, slower moving waters, forming mats of tiny, bright-green rounded leaves on the surface; both will grow in sun or shade. Propagation cannot be prevented once you have either of these species, so keep their weediness in mind before introducing them near natural waters. (Callitrichaceae)

*Caltha* Marsh Marigolds produce tussocks of kidney-shaped leaves in shallow water (to 15 cm deep) or moist to wet soils. None of them likes too much heat or sunlight, and they will die back quickly in summer if not suitably protected. The glossy-yellow spring flowers are typical buttercups, larger and showier than most. Two forms of European Marsh Marigold (*C. palustris*) are available in Australia.

*Caltha palustris*
'Flore Plena'

These are the wild plant and a double-petalled form called 'Flore Plena' with pom-pom flowers resembling some of the more compact dahlias. The giant Himalayan Marsh Marigold (*C. polypetala*) is sometimes regarded as a subspecies of *C. palustris*, but has big dramatic leaves to nearly 30 cm across; the flowers are much larger as well (to 7 cm across) and very showy. *C. polypetala* must be grown in water or really boggy soil if it is to reach full potential, but *C. palustris* will also do quite well in moist garden soils. Plantlets form on the flowerstems and can be pinched off and replanted; older plants can be divided. (Ranunculaceae)

*Camassia* Marshy meadows of northern America are the home of several species of these edible bulbs which should be allowed to dry out after flowering. All species will grow in very wet soils while in leaf, and *C. quamash* (usually still sold as as *C. esculenta*) will get by even in very shallow water. This species has columns of medium-blue flowers up to 30 cm, while *C. leichtlinii* may produce flowerspikes up to 1 m high, and comes in a number of colour forms. White and various shades of blue are available, as is a creamy-green double-petalled form which is unappealing. Maximum sunlight will give best growth and flowering, and they can be propagated by removing offset bulbs or from seed. (Liliaceae)

*Canna* The common hybrid Canna comes in a large range of colours and patterns in both flower and leaf, and in varying heights too. All forms I have tested will grow vigorously in wet soils, and most will do well even in waterlogged soils in full sun. However, as they grow perfectly well in any well-watered garden bed it would seem to be a waste to use valuable pond space for these relatively large plants unless they are a part of a special planting scheme. A wild species with smaller flowers has been grown as Queensland Arrowroot (*C. edulis*), and is one of the commercial sources of the fine-grained, easily digested starch called arrowroot. Propagation is by division, but Arrowroot will come true from seed if not pollinated by hybrid forms. (Cannaceae)

*Cardamine* Bitter Cress (*C. lyrata*) is a small-leafed, rather weedy creeper for shallow water, and is not attractive enough to be worth the space it will quickly occupy. Any piece will put out roots during the growing season while floating in water. (Brassicaceae)

*Carex* This widespread genus of rather grassy-looking sedges is usually associated with water. Several New Zealand species have unusual colouration. 'Frosted Curls' (*C. albescens*) forms slowly spreading tussocky clumps of slender, curled, silvery-green leaves. *C. comans* is similar in appearance, but the common form is a plain green; there is also a reddish-bronze form which is still rare in cultivation. *C. flagellifera* is more upright in habit and looks a little as though it has been dusted with pale, powdered chocolate.

*Carex paludosa*

The species described above like wet soils and sun, tolerating boggy conditions for weeks at a time, but *C. petrei* needs drier conditions and should never be grown in soils which are more than wet. *C. petrei* is a glossy orange-bronze version of *C. flagellifera*, and has been confused with *C. buchanani* in Australia.

The New Zealand species form a distinctive group of their own, but other attractive species come from all over the world. *C. morrowii* 'Aureovariegata' is a wider-leafed form with broad white stripes suited to very wet or moist, shaded positions; however, the following three forms described prefer shallow water or waterlogged soils. Weeping sedge (*C. pendula*) bears slender, hanging tassels at regular intervals along the flowerstalk, though the plant itself looks like a broad-leafed grass; this species has shown some potential as a weed and should not be planted near natural waters. Blue sedge (*C. riparia*, 'Coerulea', though most commonly sold under the old name of *C. paludosa*) is broadleafed and very upright; in sunlit positions it is coloured a distinctive dusty-blue, particularly in spring. Although the foliage alone makes this species worth growing, this form also produces furry, jet-black flowerheads which turn bright yellow when the flowers first open.

Bowle's Golden Sedge is generally referred to as *C. riparia* 'Aurea', but the flowers I have seen look much more like those of *C. elata*. This is the best-known *Carex* of all, with bright, golden-yellow foliage that contrasts beautifully with dark waters, though it needs some protection from the hottest summer sun. On the other hand, it will become greener in shady conditions (especially in winter), and a compromise between the two sets of conditions is hard to achieve; it may be best to keep it potted and move it around with the seasons. All *Carex* forms are easily propagated by division, and some also set fertile seed. (Cyperaceae)

*Chimonobambusa* Two species of this genus of bamboos are available in Australia. The leaves of both look best with some shade, though with full sun the stems of Marbled Bamboo (*C. marmorea*) turn a glossy purple-black quite distinct from that of the true Black Bamboo (*Phyllostachys nigra*). As this is a slender species reaching only 3 m in height, it has potential as an attractive hedge or screen in any wet soils and will even tolerate some flooding as long as the water is always moving. Although it is a moderately invasive species, it can easily be controlled by simply mowing or stepping on and snapping new shoots wherever they are not wanted — once the growing season is past it will not try to spread further for the rest of the year. There is a smaller form with white variegations on the leaves, but it is not exceptional. The other species is Square-stem Bamboo (*C. quadrangularis*); older plants reach twice the height of Marbled Bamboo, and their stems are indeed almost square. This is not as fast-moving as *C. marmorea* and will not tolerate waterlogging. A variegated-stem form has been imported but has yet to be commercially released. Both species are propagated by division. (Poaceae)

*Chondropetalum* Only one species (*C. tectorum*) is in cultivation in Australia, an erect, dark-green rush with broad, papery, chocolate-brown sheaths at each node which give it a banded appearance. This is a plant with weed potential as it sets seed freely, but the problem can be avoided by growing only male or female plants. Any moist to fairly wet soil is suitable, and propagation is by seed or division. (Restionaceae)

*Colocasia* Taro (*C. esculenta*) is possibly the oldest of all cultivated crops; it is believed that rice may have first been noticed growing as a weed in Taro fields, and then graduated to the status of an important crop in its own right. Taro comes in hundreds of varieties of which relatively few are available in Australia. Those available are mostly un-named although many are good-quality varieties, and some have been growing here for centuries (Taro is so widespread in the north of Australia that it has even been regarded as a native plant).

Tubers and leaves are edible in many forms, but make sure you are growing an appropriate variety and not one grown for ornament alone. The most commonly sold cultivar in Australia is the richly coloured, purple-veined variety which is not pleasant to eat. Another strikingly ornamental form looks to be made of maroon-black velvet; there are also several other highly ornamental forms in cultivation. Ornamental or not, any tuber or leaf on *any* Taro must be cooked properly before eating, otherwise you will have a swollen, irritable mouth which may reoccur for months.

Taro is best grown in moist to wet conditions in southern Australia although many varieties will tolerate waterlogging and shallow water over the crown during the warmer months. Plants in southern areas grow more slowly than they would in the tropics and should have full sun for maximum growth-rate or foliage colour. It isn't possible to list specific varieties at present, but many of the edible varieties I have tested have large, central tubers, and smaller tubers at the side which can be broken off for propagation. When the main tuber is harvested, the top can also be cut off with the stems and used for replanting. (Araceae)

**Cortaderia** All Pampas Grass species grow well near water; however, they are serious potential weeds which have been declared noxious in some states, and are likely to be banned in all of southern Australia before long. Many spread quickly by seed and are very hard to destroy; so don't grow any unless you wish to spend years destroying this fast-spreading weed. (Poaceae)

**Cotula** Waterbutton (*C. coronipifolia*) was regarded as a native for over a century — an indication of how quickly it established and spread itself from seed in Australia. The fleshy, much-divided leaves are almost hidden beneath yellow button-flowers for most of the warmer months if grown in full sun, but this species will flower reasonably well even in shade. Propagation is easy from seed or by division, but bear in mind that once established it's almost impossible to eradicate. (Asteraceae)

**Crinum** Many members of this genus are water-loving, and some are truly aquatic. None of the permanently submerged species has been available here to my knowledge, and in any event they are all quite tropical in their temperature requirements. *C. campanulatum* is a southern African species which grows happily underwater in the summer months, but should be kept drier in winter without ever being really dry. Pot culture is obviously the easiest way of providing these conditions — just lift the pot out of water in late autumn and let the winter and spring rains take care of the watering until it is ready to replant with the onset of warmer weather. The pinkish, red-striped flowers are small for a *Crinum*, though a reasonable number are produced. *C. bulbispermum* is a much larger plant for the water's edge (keep the bulb itself above permanent waterlevel) with large pink or white flowers. Another species, *C. capense*, has yet to flower for me, but sits sullenly in its pot dreaming of warmer climes. Bulbs of all species should be planted in shallow water in very sunny locations, and are propagated by division once they have formed sufficiently large clumps. All are slow to establish, resent unnecessary disturbance and can be shy-flowering in cooler climates. (Liliaceae)

**Cypella** These unusual bulbs from southern South America are naturally associated with marshy places and will grow in any wet conditions short of permanent waterlogging. The flowers are rather iris-like and each one lasts just a single day; however, new flowers appear to replace them regularly over a long season. Full sun and constant moisture are essential, and propagation is usually by separation of the bulbs, though seed may also be viable. *C. herbertii* produces wavy-edged, orange-yellow flowers in large numbers; another form (*C. herbertii brevicaulis*) holds canary-yellow flowers closer to the ground. *C. platensis* (also sold as *Herbertia platensis*) is a sky-blue variation with a hint of lilac. (Iridaceae)

*Cyperus* The general features of this genus have been described with the native *Cyperus* species, but the group is widespread and many attractive non-native species are also grown. Most species will grow in shallow water or permanently moist to wet soils and should have as much sun as possible — exceptions are mentioned below. *C. alternifolius* (still frequently sold as *C. involucratus*) is among the best-known species as it is frequently used as an indoor plant. Neat, flattened rosettes of rib-like leaves at the top of tall stems give it the common name of umbrella-rush. The typical form may reach over 1.5 m, but a dwarf variety between 30 and 60 cm is also available. *C. flabelliformis* is sometimes regarded as a form of *C. alternifolius* and is mainly distinguished by the broader, fleshier leaves.

*Cyperus haspan*

*C. papyrus* is also widely familiar, a very tall plant reaching 3 m in warm shallow water, but usually much smaller, especially when grown in just well-watered conditions. The stems droop slightly and are topped with masses of fine, almost hair-like foliage. This plant is famous as the raw material for paper-making in ancient Egypt — paper made from it has a distinctively smooth yet interestingly patterned texture — and the rhizomes are edible. The plant offered as Dwarf Papyrus is a different species (*C. haspan*) reaching only 45 to 60 cm high. Although attractive, it is more erect in all parts and lacks something of the soft look of *C. papyrus*.

*C. acutus* is worth growing for the contrast between its richly coloured, dark-green leaves and creamy flowerheads. *C. diffusus* is a papery-leaved form which needs shady conditions and moist soils, and which (unlike most *Cyperus*) will not tolerate any waterlogging. The common form is dark green, though a cream-striped, variegated form is sometimes available. The last species to be mentioned is *C. eragrostis* (Drain Sedge), which is attractive for its bright, pale-green flowerheads, but is a fast-spreading weed. Unfortunately, Drain Sedge has been sold in ordinary nurseries under a number of incorrect names including those of native species. (Cyperaceae)

**Cyrtosperma** Like many aroids, Swamp Taro (*C. chamissonis*) is a bold, large-leaved foliage plant worth growing for those characteristics alone. It produces edible tubers, though of quite poor quality — an early white settler in Micronesia commented that they were eaten mashed with butter or in 'cheesy, steamed slabs', but were incredibly indigestible either way. While this description reflects European prejudices of the time, a taste for Swamp Taro is probably not easy to acquire. It is mostly grown in places where water is brackish or otherwise unsuited to any other crop, and requires quite a bit of preparation and cooking before becoming acceptable as food. It will grow in diluted seawater and the hottest sun, but is not a cold-loving species and is probably best treated as an unusual ornamental in southern Australia. (Araceae)

**Didiplis** Water Hedge (*D. linearis*, also still called by an old name, *D. diandra*) is a finely-leaved, submerged plant I have a particular fondness for, even though it is not spectacular. The closely paired leaves form flattened chains along the brittle stems and in good light are strongly flushed with red. Although it will grow in full sun, it is shade-tolerant to a degree and makes an excellent and distinctive aquarium plant in water which is not too warm. Cuttings will take root, but it is better divided. (Lythraceae)

**Dietes** Butterfly Flag (*D. bicolor*, also available under an old synonym, *Moraea bicolor*) is an elegant rather than bright beauty, with flattened, pastel-yellow flowers marked at the base with a deep-brown blotch, giving the appearance of a butterfly. It is not a waterplant, yet it will grow in a wide range of conditions from very dry to very wet and will even tolerate waterlogging for a few days at a time. Full sun gives best results, but semi-shade is also tolerated. Propagation by division is easy, though the plant may take some years to settle down to flowering again. (Iridaceae)

Dionaea muscipula

*Dionaea* The Venus Flytrap (*D. muscipula*) is perhaps the best-known of all carnivorous plants, with toothed leaves that close rapidly when tiny trigger-hairs on their inside surfaces are touched. While harmless to anything much larger than a medium-sized insect, it looks very dramatic. What is not generally known is that this is a true bogplant that requires full sun and saturated soils, and will even tolerate prolonged flooding. The main reason the Venus Flytrap has a reputation for delicacy is simply that many people regard it as exotic and to be treated with kid gloves, whereas in fact, shade, warm temperatures in winter and a 'normal' watering regime are disastrous for this plant.

Many elaborate soil mixes have been prescribed for this species, but the easiest is half-and-half fine sand and peatmoss. In both cases make sure these are free of salt — sea sand and some types of peatmoss are quite saline or alkaline. On no account should this (or any other hardy, carnivorous plant) be fed. The Venus Flytrap is perfectly adapted to the acidic, nutrient-poor peat swamps of northern America, which is why they are so efficient at trapping insects as a source of nutrients. Propagation is by reasonably-fresh seed (if you can get it) or division of older, multi-crowned plants. The main problems experienced in their cultivation are their need for soft water, and smaller plants may sometimes be swamped by the rank growth of some mosses which enjoy similar conditions. (Droseraceae)

*Echinodorus* Swordplants are best known as foliage-plants for aquariums, requiring tropical conditions, but a number of species come from more temperate parts of South America. Grown in a pond, some will form remarkable specimens which rear far out of the water, while others will make uniform lawns beneath the surface. As you would expect from an aquarium plant, they are both shade-tolerant yet undisturbed by the hottest sun if gradually acclimatised to pond life in spring or autumn. *E. argentinensis* has rounded, rather featureless leaves which do not compare in dramatic effect to the related but larger-growing *E. cordifolius* (still sold in Australia as *E. radicans*) which may reach 1.5 m in height above water if grown in the shallows. The heart-shaped leaves are striking, and later in the growing season columns of conspicuous white flowers are produced.

*E. osiris* tends to keep most leaves below water and prefers being in deeper water. The leaves are semi-translucent and very lush-looking but are inconspicuous in most ponds. A red form (*E. osiris* 'Rubra') seems to show more leaves above the water, but it could be that they are just more eye-catching as they become a deep, dull-red with the onset of cooler weather. *E. latifolius* is not showy, yet will quickly form a dense and uniform underwater lawn, usually about 15 to 20 cm high. Grown in very shallow water or on waterlogged soils it is rather straggly and uninteresting. (Alismataceae)

***Egeria*** Formerly regarded as a species of *Elodea*, *E. densa* is an attractive submerged plant similar to the closely related *Elodea canadensis*. Both have small, translucent leaves that form whorls around the brittle stems, and are frequently offered as exceptional 'oxygenators', a myth which is debunked in Chapter 1. They are very adaptable to a wide range of conditions and grow so freely that they are inexpensive — in other words they are weeds, *E. canadensis* being perhaps the worst of the two. Despite having some specific uses in biology classes, they should only be grown where there is no danger of them escaping into natural waters.

Another close relative is the more curly-leaved *Lagarosiphon major* which has not yet escaped into the wild in this country but is potentially disastrous if it should do so. This species will grow in very deep water and has the potential of clogging large irrigation and hydro-electric dams; as with all such serious weeds it would cost a lot of money to control, and if you pay taxes you know just who pays. *Lagarosiphon* is still being offered under the counter by one watergarden nursery with the claim that it is only sold to 'responsible people'. What responsible people would want from this not particularly interesting weed is difficult to imagine. (Hydrocharitaceae)

***Eichhornia*** Water Hyacinth (*E. crassipes*) is the only species of this genus in Australia at the present time, and is a serious weed world-wide. Despite this, it is widely grown as an ornamental. It is certainly an exceptionally attractive plant; the strangely swollen stems keep the glossy, rather heart-shaped leaves afloat, and columns of fragrant, mauve flowers appear for months in warmer weather. However, in suitable conditions it grows at quite unbelievable rates, sets seed freely (these can lie dormant for 15 years), provides excellent mosquito habitat, and can choke the largest rivers with dense rafts of vegetation you can almost walk on.

For some reason, Water Hyacinth has been freely available from a number of nurseries in the past. Many people find this plant so attractive that they don't seem to care that possession can attract heavy fines. For this reason, I offer the following advice on keeping this plant captive. Firstly, destroy every plant surplus to your needs by composting or drying out in full sunlight; it looks most attractive grown in a group of two or three plants rather than a mosquito-ridden thicket covering your pond and suffocating any other underwater life. Secondly, cut off every flowerstalk as soon as it has finished flowering; this will prevent seed forming but is impossible to do if you have already let it slip beyond control. Thirdly, don't pass it on to anyone who doesn't fully understand why it is important to keep under control in this way. Finally, should you attract official attention, keep in mind that your only hope of not being hung, drawn and quartered is to be able to show that you have been absolutely consistent in keeping it under control. (Pontederiaceae)

***Elegia*** The only species available is the very beautiful *E. capensis*, a bamboo-like plant for moist to wet soils which receive a fair amount of sun. Masses of slender, hair-like tassels hang from each node, forming neatly symmetrical curtains reminiscent of a slender long-leafed pine tree. Propagation is by division or fresh seed if you can get it. (Restionaceae)

***Eleocharis*** The general features of Spikerushes have already been discussed with the native species. Of the few non-native species cultivated in Australia, only *E. vivipara* is so weedy that it should not be grown. Although it is unlikely to become a serious problem, it forms a dull, coarse and fast-running tangle which has little to recommend it as a garden plant. *E. acicularis* (Hairgrass) is much finer and smaller-growing, but is best used as a delicate lawn submerged in an aquarium. Grown out of water, it browns off in winter and can look very untidy.

The most important exotic species (although wild forms are also native to northern Australia) is *E. dulcis*, the Chinese Waterchestnut. The sweet and crunchy tubers can be eaten raw or cooked, but are usually only tasted in Australia in canned form. There are probably several clones in Australia of which Hon Matai is the only named one. Other forms of varying quality have appeared over the last seven years and seem to have been smuggled in by ignorant people who haven't bothered to find out that the best forms have already been imported legally. These other varieties have arrived without names, and it is likely that fully documented importations will supersede them with time.

*Elegia capensis*

*E. dulcis* can be one of the most productive plants known, with yields up to 50 tonnes per hectare; however, it requires elaborate treatments to achieve anything like the maximum figures possible. A long and reasonably warm growing season of at least seven months is important, otherwise plants should be started earlier under glass in 12–15 cm diameter pots. Whether planted direct or in pots, the tubers should be kept wet but not waterlogged until new growth is at least 15 cm high. At this stage they are ready to be permanently placed allowing three plants for every two square metres of surface area; greater densities will reduce yields. The pond should then be flooded to a depth of 7–10 cm which must be consistently maintained throughout the growing season.

Full sun and fairly open soils are also important, as shade or a heavily clay-based mix will inhibit tuber formation. Although the soil should be rich in organic matter and nutrients, too much nitrogen will encourage top growth at the expense of tubers; the addition of lime is often necessary. During autumn the foliage will start to yellow, a sign that the plant is starting to put energy into the future tubers. Drain the pond to encourage tubers to form, but don't let it ever dry out completely. After another two months or so the foliage will be completely dead, and at some time in July or August the tubers will be ready to harvest. Digging up the tubers is easy enough, though commercial crops are most easily gathered with a modified *Gladiolus*-harvester once the soil is sufficiently dry. If you can't provide the conditions described above but would like the crisp texture of waterchestnuts in your cooking, it may be worth trying Kuwai (*Sagittaria sagittifolia*), described later in this chapter. Another plant called Waterchestnut is discussed under *Trapa*. (Cyperaceae)

*Elodea canadensis* is discussed with *Egeria*.

*Equisetum* This alphabetical listing is starting to read like a catalogue of weeds, and here are four more potential ones. Horsetails are fascinating and strange rush-like plants with an ancient history, but in the northern hemisphere many are invasive and persistent weeds. All species are indifferent to almost every herbicide invented, and those few herbicides which will eradicate Horsetails are so dangerous it isn't even safe to read their labels. However, in order to grow *Equisetum* so that they look their best, choose a site with some shade from the hottest sun. The species grown in Australia enjoy moist to wet soils and will tolerate flooding for short periods of time; all species are probably best grown (and safely contained) as potplants stood in a shallow tray of water.

In the early stages of the colonisation of land by plants, some Horsetails reached the size of trees; however, the survivors are much smaller, all being less than 2 m high, and none in this country being much more than half this height. *E. hyemale* is the tallest, with simple, dark-green, ridged stems neatly marked black at the joints. It is the only species actually established as a weed at present (forming dense stands where it is not wanted in the Geelong Botanic Gardens). A species similar to (but not necessarily identical with) *E. arvense* is very distinctive, forming colonies reminiscent of pale-green Christmas trees less than 20 cm high. If it is the true *E. arvense*, it has been used in the past by

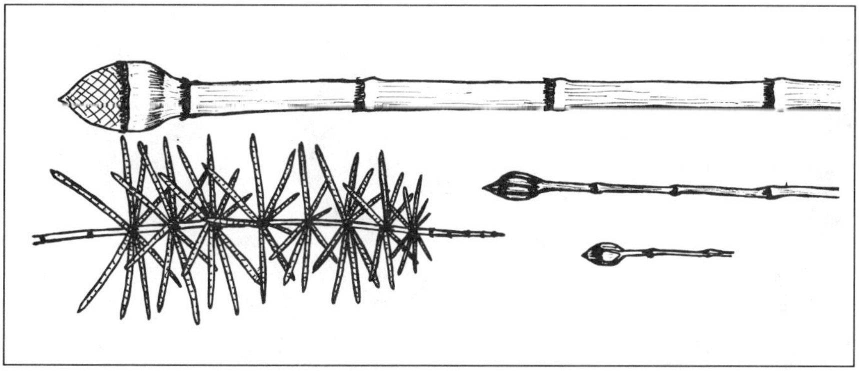

*Equisetum hyemale* (top), *E. arvense* (bottom left), *E. scorpoides* (centre right), *E. variegatum* (bottom right)

herbalists and is an important component of some biodynamic preparations. Unlike the other species, this one is a favourite with slugs and snails. If *E. arvense* is kept in pots as described above, control of both plant and pest is easy as neither will cross the water-barrier in the tray the pot is stood in.

The remaining two species are similar in general appearance — but *E. scorpoides* is a tiny miniature up to 5 cm high, while *E. variegatum* forms much taller thickets to 30 cm high. Both are graceful in appearance with slender, wiry stems ending in small, unusually ridged cones. (Equisetaceae)

**Eriophorum** Cottongrass (*E. angustifolia*) is a low-growing, deep-green sedge which spreads rapidly in shallow water and on waterlogged soils. The flowers are inconspicuous but are followed by very showy tassels of pure-white cotton. At best, Cottongrass may form a dazzling white carpet almost too bright to look at in intense sunlight which also is best for promoting flowering. Although I know of no cases where this species has spread into the wild in Australia, it does seem to produce a fair amount of wind-blown seed and should not be planted near natural waters. (Cyperaceae)

**Eupatorium** The North American Joe-Pye-Weed (*E. purpureum*) is a very tall and coarse-looking herb topped with flattened, purplish flowerheads. European Hemp Agrimony (*E. cannabinum*) forms a smaller plant with finer and more delicate foliage. The wild form of Hemp Agrimony is pink-flowered, but I have only seen the white-flowered

*Eriophorum angustifolia*

variety grown in this country. Both species need moist to wet soils and will not tolerate flooding for more than a few days; some shade from the hottest sun is necessary to prevent the foliage yellowing in hot summers. (Asteraceae)

*Euphorbia* Swamp Spurge (*E. palustris*) is a rather non-descript plant resembling some fast-spreading, weedy species of this genus with thin papery leaves. It hasn't been available long and may have moderate potential as a weed; keep this in mind before you turn it loose where it can spread too freely. It will grow in any moist to waterlogged soils, and even in shallow water. In summer, the bushy plants are hidden under sulphur-yellow flower bracts; in autumn the leaves give a fine show of orange and reddish colours. Full sun or light shade are preferred, and propagation is from seed. Division is probably possible also, but cut plants freely produce a poisonous white sap. (Euphorbiaceae)

*Filipendula* At least six species or forms of this genus are available in Australia although, strictly speaking, some may now belong to other genera. All varieties are waterside plants requiring permanently moist to wet soils; they will tolerate some flooding but will be weakened if submerged for any length of time. Full sun gives good growth and flowering; some shade may be needed in hotter areas. Most species have finely-divided leaves in various shades of dark green (often with purple or red tints in selected garden forms), but only *F. hexapetala* is really distinctive — the finely and repeatedly divided leaves of this species form flattened rosettes somewhere between moss-curled parsley and some ferns in appearance. Flower-columns in all species are similar to those of *Astilbe*, yet are flattened rather than column-forming. (Rosaceae)

*Fritillaria* Only one species of this large genus of lily-like plants is associated with water, the Snakeshead Fritillary (*F. meleagris*). In nature this grows in watermeadows, basically riverflats which are always moist and may be flooded in spring. The bulbs themselves must always be above permanently wet soil, and growing plants look best when shaded from the hottest sun. As a passionate admirer of *Fritillaria* generally, I must be careful not to wax too enthusiastically. Suffice it to say that this is one of the most beautiful, inexpensive and easy-to-grow species. The elegant bell-shaped flowers vary in colour from deep purple to reddish pink or almost white, and are usually patterned with a muted, paler chequering. A number of named varieties with larger bells and distinctive colours exist but few are available in Australia, and the Snakeshead Fritillary is usually sold here as assorted, mixed-colour bulbs. (Liliaceae)

*Gentiana* Although all Gentians love permanently moist conditions, none do well in really wet or waterlogged soils. Willow Gentian (*G. asclepiadea*) with its unusually tall (to 60 cm) stems and pendulous, violet flowers is reputed to do well at the edge of ponds, but has grown poorly for me. Many species require shade in Australian conditions and I suspect that their poor growth is more related to excessive heat than to the wetness of their soil. Other particularly wet-loving species are available occasionally, though they don't like warmer climates or lower altitudes, which is to say most of Australia. (Gentianaceae)

**Geum** As grown in this country, Water Avens (*G. rivale*) is a low-growing, herbaceous plant with small, nodding pinkish flowers, lacking the red stems and richer colours of the plant known by this name overseas (the true *G. rivularis* may be available in the near future). It is possibly of hybrid origin and is more usually grown as a cottage garden plant because it prefers moist rather than really wet soils. Although easy to grow and tolerant of sun or semi-shade, I can think of many more interesting plants for the watergarden and have included it here for the reason that it is available and is frequently listed in European watergardening books. (Rosaceae)

**Gladiolus** A number of *Gladiolus* species occur naturally in sunny, open areas which are very wet or even waterlogged for all of their growing season, but need dryer soil once the foliage has died down. *G. tristis* is a distinctive, narrow-leaved species with flowers which are very fragrant in overcast weather (or at dusk and dawn). The common form of this species is creamy-green marked with reddish brown in the throat, but there is also a deceptively-named unmarked form of similar colour called 'Alba'. In some areas it is a declared noxious weed as the foliage is poisonous to cattle and the plant will spread quickly from seed given suitable conditions.

The other swamp-loving species have more typically broad, *Gladiolus* foliage, but their flowers are all different from the rather bland shape of the hybridised commercial cultivars. *G. palustris* has magenta-pink flowers, while *G. byzantinus* has rather bunched-up flowers of a rich and velvety reddish-purple — the throat is marked with white blazes on a pinkish-purple background. *G. undulatus* should have creamy-coloured flowers with rather crinkled petals, but there appears to be some confusion with *G. carneus* and hybrids between the two species, so more than one type may be offered under this name. (Iridaceae)

**Gunnera** The non-native members of this genus can be almost unbelievably different in appearance, yet their cultivation is similar. All grow best with full sun and wet soils, with only the roots growing into the permanently waterlogged zone. *G. hamiltoni* is a creeper with rounded, veined, rather purplish leaves. After flowering, it produces columns of bright-red berries held well above the foliage, making the plant quite showy despite the small size. Many other *Gunnera* species are small-growing with interestingly arranged leaves and fruits; examples including the creeping *G. dentata* with tiny oak leaves or *G. perpense* with big kidney-shaped leaves held high above the crown.

Chilean Rhubarb (*G. manicata*) is a very different plant. The leaves of this species are immense and rhubarb-like (though it is not edible), held on stout spiky stems. At its best, a leaf may reach close to 3 m long and more than half as wide, therefore Chilean Rhubarb is not recommended for smaller gardens. As you would expect from such a huge plant, it is a gross feeder, requiring hefty servings of well-rotted animal manure each spring — add blood-and-bone if you use cow manure. The very similar *G. chilensis* may be synonymous but it is slightly smaller and reputedly hardier; neither form needs winter protection in any Australian climate much below the snowline. *G. manicata* has been advertised recently as 'Giant Amazon Perennial' at remarkably high prices. Those in the know buy identical plants from their local garden centre for a fraction of the price. (Haloragidaceae, syn. Haloragaceae)

*Gunnera manicata*

---

*Gymnocoronis* Senegal Teaplant (*G. spilanthoides*) is widely available through the aquarium trade. It is a drab herby-looking thing which has distinguished itself mainly by showing very definite signs of weed potential in warmer climates (even though it is quite hardy almost everywhere in Australia). The creamy flowerballs are only slightly more interesting than the foliage, and it will grow in full sun or shade, from deep water to waterlogged soil. It can be propagated from seed or division and any piece broken off will also take root. (Asteraceae)

***Hemerocallis*** Daylilies are too well-known to need description here; it is enough to say that almost every species and hybrid grows best in full sun with constant moisture, or in waterlogged soil. Some may even grow in shallow water, though these don't seem to grow as quickly or flower as well as those planted at the water's edge.

This is a popular group of garden plants with hundreds of varieties available at any time, and dozens of new ones being released each year. A number of nurseries specialise in Daylilies, alone or in combination with irises; their current catalogues are the best source of information and varieties for this genus. Most parts of a daylily are edible, and so surplus plants are easily disposed of. (Liliaceae)

***Heteranthera*** *H. reniformis* can be very fast-spreading and weedy, especially in warmer climates, but frosts will set it back temporarily. This species forms a fascinating carpet of kidney-shaped leaves and has small, deep-blue flowers which are attractive but not showy. Although it will tolerate shade, best growth and flowering is achieved in full sun in shallow to moderately deep water. Propagation is easy — obtain a piece and turn it loose. (Pontederiaceae)

***Hibiscus*** Only one *Hibiscus* in Australia is truly aquatic, the spectacular Swamp Hibiscus (*H. coccineus*), with big blood-red, glossy flowers. The waxy pink stems reach 2 m high and bear serrated leaflets in the form of a hand (I have even had it mistaken for marijuana!). Although the entire plant dies back with cold weather it will regrow rapidly in late spring in all but the coldest climates. Full sun and constant moisture are essential, with even shallow water being tolerated. Propagate by seed or division. (Malvaceae)

***Hosta*** Plantain lilies are usually grown for their foliage, however, they also produce columns of somewhat lily-like flowers in shades of purple, lilac and white. They have been attracting considerable interest in recent years — both here and overseas — and some very attractive cultivars are now available. Hostas are really woodland plants with a high requirement for water, though they will not tolerate waterlogging. Semi-shade is necessary for many varieties to achieve their richest colour and best development of foliage. Others will look almost as good with a considerable amount of sun as long as they remain constantly wet; and some will grow in very deep shade.

Entire books have been written on this genus; however, most varieties in cultivation are not yet available in this country. Nevertheless, there are probably at least 50 here, so I will only give a brief description of the few I particularly like.

A number of wild species are still available, but many don't look quite as interesting alongside the more impressive hybrids. *H. fortunei* is an example of these superseded beauties. It is a fairly average green and

*Hosta undulata* 'Medio-variegata'

has elegantly sculpted leaves. There is also a yellow-variegated form of *H. fortunei* ('Aureomarginata') which doesn't maintain the full colour of the variegation as the foliage grows older. *H. lancifolia* has elongated, undulating leaves while its variegated form ('Aurea') also becomes more uniform in colour as the growing season progresses.

*H. sieboldiana* has withstood the test of time with almost round leaves which are corrugated and pitted, and are a dusty blue-grey in spring. Unfortunately, random seedlings of this species are often offered which rarely show the perfection of form of the best plants. *H. plantaginea* is a big and bright-green species which would stand out in the company of any foliage plants. The variety 'Grandiflora' has particularly large, fragrant white flowers. *H. undulata* is represented only by variegated forms, all of which look very similar. This is not a criticism however, as they are among the best variegated Hostas I have seen, with the centres of the leaves marbled in pure white and various shades of green. The variety most commonly available is 'Medio-variegata' which holds the variegation very well through the whole growing-season.

More recent variegates are complex hybrids, and generally also hold the colour of the variegation quite well. 'Francee' has won much praise, though it is not really showy to my eye; the bluish leaves are edged with a slender border of near-white. 'Francis Williams' is possibly my favourite variegated *Hosta*. It has very rounded, near-blue leaves held flat and low. These are distinctly corrugated and pitted, and bordered with a creamy-yellow band which contrasts nicely with the central colour without being too different from it.

More recent, non-variegated varieties are sometimes even more striking than the variegates. 'Big Daddy' is a particularly large-leafed variety of deep blue-green, neatly corrugated, free-flowering and fast-growing. 'Zounds' is not a large plant and is almost a metallic yellow which seems to glow at twilight. Grown in shaded conditions it becomes much greener.

Not all *Hostas* make good specimen plants; in fact, there is a whole range (called edging varieties) bred to form dense, low groupings which make an unusual border in shady areas. The colour range is similar to that seen in the larger varieties, but the foliage is often plainer in texture. Although I appreciate the rich blues of some edging varieties such as 'Blue Wedgwood', the yellows, such as 'Honey Bells', seem quite insipid to me, rather like bleached plantains. If you are planning to become a *Hosta* enthusiast, invest in one of the many books which have recently been published on this group. (Liliaceae)

*Houttuynia* Polypora (*H. cordata*) is a densely growing creeping plant with smallish, heart-shaped leaves edged red. It needs constantly moist conditions but is happier when waterlogged or even in shallow water, and will grow in full sun or quite deep shade. The common form can be fairly invasive in suitably wet locations, and there is also a double-flowered form which looks very similar until the much more showy flowers appear. A much less invasive form is 'Variegata' (also called 'Chameleon' or 'Court Jester') with extraordinary, multi-coloured leaves in varying shades of red, green and creamy yellow, all splashed with silvery-grey. All forms are easily propagated by division. (Saururaceae)

*Hydrocleys* Water poppy (*H. nymphoides*) is an attractive flowering plant for deep or shallow water, but is a potential weed in warmer climates. The floating leaves are spoon-shaped, forming dense tangles quickly in hot weather. Flowers are quite large and showy — a delicate yellow with prominent dark markings at the base of each of the three petals. Although some shade is tolerated, full sun is best for flowering. New plants can be propagated from any rooted piece along the runners forming on established plants. (Limnocharitaceae)

*Hydrocotyle* *H. bonariensis* is sometimes grown as an unusual foliage plant for its rather shield-like leaves. This is a tenacious weed which will survive permanent flooding or complete drought once it is established. Any piece with a node will take root in wet conditions and full sun, but shade is also tolerated. (Apiaceae)

*Hymenocallis* Many hybrids and several species of Spider-lilies are grown in Australia (all fairly similar in general appearance), of which two species are naturally associated with water. *H. littoralis* has large, white flowers resembling daffodils with slender threads drawn across them. Although it will grow in full sun or with some shade, this is not an easy species to flower in cooler climates and can be slow to establish; constant moisture is essential, and flooding for a few days at a time doesn't seem

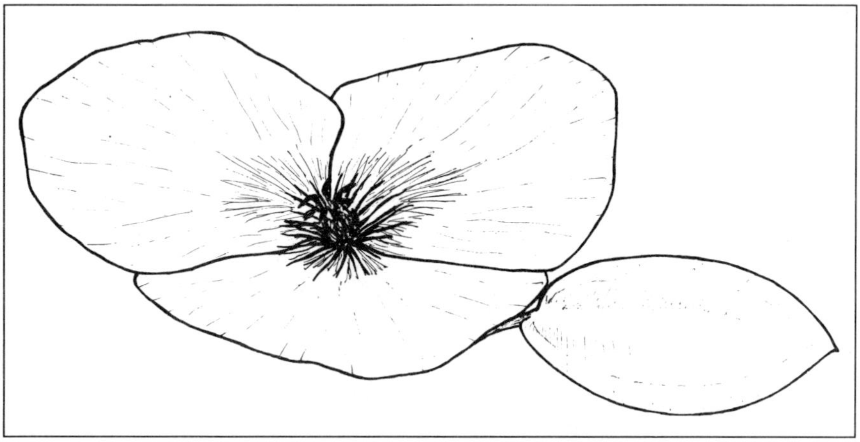

Water Poppy *Hydrocleys nymphoides,* a popular non-native ornamental with considerable potential as a weed in warmer climates. The native *Ottelia ovalifolia* is similar in general appearance (with equally large flowers and larger, more ornamental leaves) and flowers both longer and at lower temperatures.

to bother the plants even in cold weather. *H. liriosme* is truly aquatic, growing in waterlogged soils or shallow water. Propagate by dividing-off bulbs from well-established clumps, but keep in mind that this may set flowering back for a few years. Fresh seed is easy to germinate. (Amaryllidiceae)

*Hypericum* *H. elodes* is a small, round-leafed, nondescript creeper for shallow water, with yellow flowers which do little to add to its interest. Full sun or shade are suitable and it is easily propagated by cuttings or division. (Clusiaceae)

*Ipomoea* Kang Kong or Water Spinach (*I. aquatica*) is one of the most productive leaf vegetables known, with yields of up to 60 tonnes of tender, young shoots per hectare being possible over a few months in ideal conditions. The semi-aquatic form with white stems is most productive and grows best in swampy conditions, even shallow water. A land-growing form with green stems is best planted in moist to wet soils, but is much less productive than the semi-aquatic form. In most of southern Australia both forms can only be grown as annuals because they are very cold-sensitive. Plants should flower and set seed after the autumn equinox, and the seed can be stored for at least two years.

The slender foliage is not particularly ornamental, but when Kang Kong flowers a close relationship to Morning Glory is obvious in the bell-shaped blossoms. Full sun is essential for best productivity, and propagation in southern Australia is by seed only unless you wish to maintain plants for cuttings in a greenhouse over winter. Kang Kong is usually eaten after being briefly cooked — for example, stir-fried. (Convolvulaceae)

# *Iris*

There are hundreds of moisture and water-loving irises, including some of the finest species and cultivars in this genus. In view of this I give only minimal information on those types which are commonly offered for ordinary garden conditions or which are available in hundreds of varieties. Unusual irises and true water-irises, on the other hand, are discussed in some detail. All named varieties must be propagated by division, but some interesting seedlings can also be raised if you have the space and time to grow them from seed.

## Irises for moist to wet conditions

There are several common groups of irises which are suitable for moist to quite wet (but never waterlogged) soils. Of these, the Evansia-type irises are really woodland species requiring reasonably shady conditions. The best-known species in Australia is *I. japonica* with smallish but exquisitely-shaped flowers with ruffled crests. The wild plant is a pastel blue marked with yellow and purple, though white-flowered and variegated-leaf forms are also grown. Damage to the leaves will prevent the plant from flowering in the following growing season, so handle mature plants with care and keep slugs and snails away. Less common Evansias include *I. wattii, I. gracilipes* and *I. tectorum*; while all of this group are quite similar in shape (less so in colour) to *I. japonica*, to my eye they are all coarser and less attractive than that species.

The Kaempferi irises are now correctly known as *I. ensata*, but as Kaempferi appears to have become their common name in general usage I will use it as such. There are hundreds of forms available in Australia, and some have been overbred to the point where they resemble large irises with wet flannels draped across them. Many of the older varieties still available are fairly colourless, yet the modern and more colourful varieties can be considerably more expensive than their plainer, old-fashioned relatives.

Although the Kaempferi irises are sold as Japanese water-irises, they are actually woodland plants which are water-tolerant during the growing season (however, there is a true Japanese water-iris, *I. laevigata*, which this common name is better reserved for). During flowering the Japanese flood the plants in order to enjoy the reflection as well as the flower itself. Kaempferi irises dislike lime and should have some protection from the hottest sun. Keep them moist to waterlogged during the growing season but remove them from wet conditions at the first sign of the foliage dying down for the winter.

Siberian iris hybrids (*I. sibirica*) are also well-known and available in dozens of varieties. Although some old types have been offered by watergarden nurseries, the best variety can be obtained from specialist iris growers. Many newer varieties are available, few of which are great improvements on the older types (being mostly just further variations on blue, purple and white). Hopefully, further importations of newer overseas varieties will include the rich almost-black, gold, and multi-coloured varieties thus far lacking in Australia. Cultivation is usually straightforward if plants are kept moist to wet — but never waterlogged. However, in permanently wet conditions, fungal infections may kill many plants unless you routinely use fungicides. As these are likely to wash into the water with disastrous results for the underwater life, I recommend that you avoid using Siberian iris other than in normal, well-watered garden beds.

Other species sometimes suggested for wet conditions include *I. orientalis* (still sold as *I. ochroleuca*), a fairly large species with whitish flowers marked yellow. Although easy to grow in most moist soils it is not really a bogplant and can drown in waterlogged conditions. *I. chrysographes* is also reputed to be suitable for wet sites, but in my experience it rots away quickly under such conditions. Plants masquerading as the bog species *I. prismatica* are available from time to time, yet are very different in colouration from the true species and seem to dislike wet conditions.

## Irises for waterlogged conditions

All three irises described here are capable of growing in shallow water, but seem happiest in very wet or waterlogged conditions. If you wish to grow them in shallow water, a depth of 5 cm of water above the crown is the maximum acceptable.

*Iris* 'Roy Davidson'

*I. tridentata* is a small-flowered species, yet the flowers are both beautifully formed and a striking, deep purple-blue colour. 'Holden Clough' is an *I. pseudacorus* cross (possibly with *I. chrysographes*) with very unusually coloured and marked flowers — rich-gold laced with an intricate network of purplish-brown. 'Roy Davidson' is a seedling of 'Holden Clough' back-crossed against its own parent *I. pseudacorus*. This variety is reputed to retain the flowers much longer than *I. pseudacorus* but the difference is not great if it exists at all; each flower is similar to a large *I. pseudacorus* prominently veined with almost-black lines.

*Iris pseudacorus* 'Flore Plena'

## Irises for waterlogged soils and shallow water

The irises covered in this section are true water-lovers and include some of the best and most sought-after varieties of this genus. Louisiana irises in particular are very popular in Australia at the present time, partly because some of the major breeders and hybridisers live in this country, and partly because they will grow and flower well in most Australian climates. Not all of the modern Louisianas are to everyone's taste; they are bred to standards in which the degree of frilling and the overall impression of each flower have more to do with the current ideal chosen by iris fanciers than with the beauty of the flower itself.

There is no possibility of listing all of the hundreds of hybrid varieties of Louisianas available in Australia; these change constantly as iris specialists drop older ones in favour of more modern varieties. Many older (and cheaper) varieties *are* less colourful and attractive than the new hybrids, but some were the pinnacle of perfection in their time and have not been improved on by the current favourites. All Louisianas are intolerant of lime and require full sun for reliable flowering. Unless you start with potted plants or well-rooted divisions, use some sand in the mix to help new roots to establish.

The wild species used to create today's Louisianas are rarely available, yet have a special charm and 'feel' which has little to do with the exaggerations which have been imposed on some of their descendants. The ancestral species most likely to be available are *I. fulva*, a narrow-petalled, rusty-red iris with an unusual shape which allows pollination by hummingbirds, and *I. brevicaulis*, with large blue flowers. *I. hexagona, I. nelsonii* and *I. giganticaerulea* are rarely available.

*I. laevigata* is a true Japanese water-iris (Kakitsubata or Blue Flag) and must be grown in water at all times. Although plants will survive in wet conditions, the minimum requirement for good growth and flowering is permanent waterlogging; established plants do even better with 5–20 cm of water above the the crown. Wild-type flowers are a rich dark blue, sometimes with purplish tints, depending to some degree on growing conditions. They are three-petalled and similar to the generally smaller flowered *I. pseudacorus* in shape. The only single-flowered *I. laevigata* forms commonly available are 'Regal', 'Variegata' and 'Semperflorens' — the latter is the closest thing to a wild type *I. laevigata* grown in this country.

Most cultivated forms in Australia are double in the sense that they have six petals, but as these are arranged neatly and symmetrically around the perimeter of the flower the effect is both simple and natural. Wild-colour *I. laevigata* forms are regarded by many writers on this subject as the finest blue irises in the world, and I would have to agree. The large rich-blue flowers are something really special and will change your idea of what blue in irises can be like.

Wild-coloured but double flowers are also found in 'Royal Cartwheel' and 'Violet Parasol' (though the latter has not been offered under that name for some time and is very similar to 'Royal Cartwheel'). 'Variegata', occasionally called 'Elegantissima', is a relatively shy-flowering variety which is mostly grown for the bold, white leaf variegation which the plant maintains for the whole of the growing season; the single flowers are a slightly paler blue than 'Royal Cartwheel' or 'Semperflorens'.

Two white varieties have been offered in Australia: 'Alba' and 'Snowdrift'. Both are identical in every way and are probably correctly 'Snowdrift', a fairly large double with mauve and yellow markings towards the centre (the true 'Alba' is single-petalled). 'Mottled Beauty' (which has been incorrectly offered as 'Albopurpurea' and 'Montrosa', both of which are obviously different even in photographs) is also a white flower, with a scattering of purplish-blue spots towards the edge of the petals, leaving a white border. 'Colchesteri' (incorrectly 'Colchesterensis') is a similar white flower with a broader and usually denser, purplish-blue band extending right to the border of the petal. Both of the last-mentioned varieties are stunning as a large clump or group. However, the

*Iris laevigata* 'Colchesteri'

individual flowers don't have the impact of blue-flowered forms of *I. laevigata*.

'Regal' is very different in colour to the other Laevigatas, being a rich, dark pinkish-purple which cannot be confused with any other water-iris. 'Rose Queen' is also distinctly coloured, a soft pink which is not a true Laevigata but a hybrid perhaps with a Kaempferi. It is invariably listed in books with *I. laevigata*, although it does not grow well in shallow water, preferring wet to waterlogged conditions instead. I have recently started growing *I. laevigata* forms from Japan; these vary dramatically from the forms already available, but the many variations are both beautiful and subtle.

*I. pseudacorus* is the showy yellow-flag iris which can be seen from a great distance when in flower. The common form is a fast-growing and adaptable plant which can be on the weedy side for it spreads very freely by seed. In order to keep it in one place seedpods should be cut off before they ripen. Although this species will grow in moist soils, it is another true aquatic; established plants will grow and flower even in water approaching 50 cm deep.

A number of other forms of *I. pseudacorus* are available. Some of these are selected forms from distinctly coloured wild populations and may come true from seed. 'Ivory' has a creamy-white flower with subtle veining which sets off the purple centre, while 'Bastardi' is a creamy-yellow variety probably identical to 'Sulphur Queen'. 'Flore Plena' has the typical, bright-yellow colouration of the common type, but has double petals which make it look rather untidy when seen close up. From a distance, however, the flower has a soft, almost fluffy look. 'Superba' (Golden Queen) is a freely flowering yellow with broader petals than most other varieties. 'Variegata' has boldly yellow-variegated leaves when these first appear in spring, yet by the time of flowering, many weeks later, these revert completely to green.

*I. versicolor* and *I. virginica* are much confused, partly because they are closely related to the extent that they have been considered synonymous. They also hybridise quite freely, and it is possible that some of the selections and forms attributed to one or the other are actually a mixture of both. In general, *I. versicolor* sends up leaves a little later than *I. virginica*, which may also be a bit taller, more upright and brighter green than *I. versicolor*. The varieties of *I. versicolor* grown in Australia are strongly coloured selections not representative of the range of colours found in wild populations, while *I. virginica* flowers tend to be more subdued in colour. Both species and their hybrids will grow in very shallow water, wet soils, or even well-watered garden beds.

I only know of three *I. versicolor* forms presently available in Australia; all are distinctive, named cultivars. 'Claret Cup' has rich, purplish-red flowers which are beautifully veined. 'Gerald Darby' is a deep blue, and the flowers are carried on unmistakable deep-purple stems — this variety may be a cross with *I. virginica*. 'Kermesiana' is a claret-purple with whitish markings. Wild *I. versicolor* are amazingly variable and with time other colour variations such as white with various colours of veining — pinks, bicolours, etc. — will probably be available.

No named varieties of *I. virginica* exist to date, so the flower types are usually listed and described when these are offered. They vary in size considerably, from 5–10 cm or more in diameter. Colours vary from cream to pale blues, lavender-pinks, violet and various shades of purple, but are mostly more pastel in effect than those seen in the three *I. versicolor* forms.

All *Iris* belong in the family Iridaceae.

***Isolepis*** *I.* 'Fairylights' is discussed under *Scirpus* later in this chapter.

***Juncus*** Only one non-native form of this genus is even remotely worth growing, as most are potentially serious weeds which spread freely by seed. The bizarre *J. effusus* 'Spiralis' has leaves which are so spiral that a tussock looks like the springs of a broken sofa. While not an elegant plant, this unique mutant has a special charm and is best grown in very shallow water or wet soils, in full sun or moderate shade. The rhizomes are as twisted as the leaves, making division a task to challenge the most skilled surgeon. (Juncaceae)

***Kniphofia*** Red-hot Pokers (even the yellow-hot ones) are water-loving plants for permanently moist to wet positions with lots of sun, but will not tolerate waterlogging for more than a few days. All selected forms and hybrids available are propagated by division, but species can be grown from seed. (Aloeaceae)

## *Lagarosiphon major* is discussed with *Egeria.*

**Leucojum** Snowflake (*L. aestivum*) is the very familiar, winter-flowering bulb with nodding, pure-white bell flowers tipped with green. It grows happily even in shallow standing water, but must be allowed to dry out when it dies down in warmer weather. Shady positions or full sun are acceptable though the growing plant must be kept wet if in full sun. Propagation is by separation of the dormant bulbs. (Liliaceae)

**Ligularia** These daisies have been shuffled back and forth from this genus to *Senecio* in recent years. Having no opinion of my own on their status I have simply listed them under the name most familiar to gardeners and under which they are usually sold. The best varieties are striking foliage plants for moist or wet, shaded positions and will not survive flooding for more than a day or two.

*L. clivorum* has large, purplish kidney-shaped leaves with reddish undersides, and holds the tight heads of orange flowers well above the leaves. *L. przewalskii* is quite different in appearance, with mid-green leaves, deeply divided and serrated — a mature clump with black flowerstems and columns of yellow flowers is a spectacular sight. *L. stenocephala* 'The Rocket' is superficially similar and just as striking, but

*Ligularia* leaf shapes: (left to right) *L. przewalskii, L. sibirica, L. clivorum*

its leaves are only divided along the edges. This variety has been confused with *L. przewalskii* in Australia — from a small distance their general appearance is identical — so you can't necessarily be sure which one is offered when they are available.

Leopard Plant (*L. tussilaginea* 'Aureo-maculata') is another distinctive type, with glossy, rounded, slightly serrated leaves handsomely spotted with gold. *L. sibirica* is a very distinctive form with pitted and finely corrugated leaves rather similar in shape to some arrowhead (*Sagittaria*) species. It is likely that most varieties of this genus sold in Australia are selected forms as they don't necessarily come true from seed, hence propagation is usually by division. (Asteraceae)

***Lilaeopsis*** *L. brasiliensis* (sold incorrectly in the aquarium trade as *L. novaezelandiae*) is a fine-leaved, grassy plant frequently used in aquaria, which forms a lawn around 5 cm high and tolerates low light conditions. Although hardy outdoors in most of Australia, it is not an ideal pond plant as it is easily overwhelmed by larger plants. However, it may succeed in clear, nutrient-poor water or on waterlogged soils without too much competition. It is easily propagated by division or from the fast-moving runners. (Apiaceae)

***Lilium*** A number of North American representatives of the true Lilies are semi-aquatic, growing in very wet conditions in spring as ice-melt water runs over them. However, they are less tolerant of flooding in Australia, possibly because water temperatures are too high for some of them. Even so, several rather similar species (all have flowers in various combinations of red, yellow and orange usually speckled with brown or black) will do well near the water's edge year round if the bulb itself is planted well above the waterlogged zone. Flooding for several days at a time does not appear to worry them during cooler weather, and full sun is best for flower-production.

*L. occidentale* is the most water-tolerant in Australian conditions; *L. vollmeri* is reputed to be the most aquatic in a natural environment yet is prone to rot here if kept wet too long. *L. superbum* and *L. canadense* are also very water-loving at home but not here. On the other hand, *L. pardalinum* is surprisingly water-tolerant for a species naturally found on well-drained but seasonally saturated slopes. All species can be propagated from seed or by breaking off scales from the bulb and growing them in less wet conditions. (Liliaceae)

***Lobelia*** I was never much of an enthusiast for bedding plants, so it came as a surprise to me to learn that most Lobelias aren't aquatic; on the other hand, everyone else seems amazed to learn that some are! Most water-loving *Lobelia* are only 60 to 120 cm high and will grow in moist or waterlogged soil, or even very shallow water. While full sun is optimal for flowering and colour all the non-African varieties are happy enough in

*Lilium occidentalis*

semi-shade. Three species are available in Australia, including *L. syphilitica* which was once used to treat the disease of the same name; the wild-type flowers are a lilac-blue and a creamy-white form is also grown. *L. sessilifolia* produces slightly shorter columns of purplish-blue flowers and may be attacked by fungal diseases in wet, cold conditions. *L. cardinalis* is a true aquatic growing as deep as 20 cm; however, this is not the plant offered in general nurseries under this name. The foliage is an unremarkable green, though intense blood-red flowers appear in succession along metre-high stems for weeks in late summer.

The other more colourful varieties are tall-growing hybrids. *L.* 'Vedrariensis' has foliage not unlike *L. cardinalis*, and the purple-blue flowers are produced at a similar time. *L.* 'Queen Victoria' is so colourful that it borders on bad taste — the beetroot-red foliage is crowned with a column of intense red flowers similar to those of *L. cardinalis* in summer, a combination so gaudy as to make it impossible to decide whether it is dazzling or just unnatural! *L.* 'Fan' is a deep rose version which is just as water-loving having been bred from the same original species.

Many giant *Lobelia* are found at high altitudes in Africa, and some are associated with marshy ground. Of these only *L. gibberoa* has been grown in Australia to date. It forms a colony of suckering plants up to 2 m high; the older stems grow flower-columns to another 2 m or more above that. Individual flowers are small and a rather plain, creamy colour, but the overall effect is dramatic. This is a suitable companion for *Gunnera manicata* in sweeping landscapes and needs full sun in wet (though not waterlogged) soils. All wild-type Lobelias can be propagated by seed or division, but hybrids and selected forms should be reproduced by division as seed will rarely come true. (Lobeliaceae)

*Ludwigia* All *Ludwigia* species are a bit on the weedy side and shouldn't be planted anywhere near natural waters. Flowers are yellow and freely produced on plants grown in full sun; some shade will not affect growth although leaves will be greener in low-light conditions. *L. arcuata* has needle-like foliage when grown submerged, but in shallow water or on wet soils forms a carpet of tiny, glossy-red leaves overlaid with small, yellow, flared flowers. *L. palustris* has much larger leaves, is less intensely coloured and has similar growing requirements — this species has already shown weed-potential by escaping into the wild in cooler climates. Willow-leaved Ludwigia (*L. longifolia*) is a small, upright bush with willowy foliage. The leaf-tips should regularly be pinched out to keep growth compact or it will become very straggly. It spreads freely by seed so should be planted only where it cannot escape; a closely-related species (*L. peruviana*) is a fast-spreading weed in warmer climates. All species are easily propagated by cuttings when actively growing. (Onagraceae)

*Lysichiton* Both species of *Lysichiton* flower before their foliage appears, and the very upright, Arum-like flowers are really eye-catching in early spring. The leaves which follow are unusually large if a bit shapeless. These plants are superbly adapted to the cold of Alaska and Siberia but will grow reasonably well in cooler parts of southern Australia when protected from really hot sun. The canary-yellow flowered *L. americanum* is sometimes referred to as Western Skunk Cabbage because of the odour produced by its flowers. *L. camtschatcensis* is slightly smaller in all parts than its relative from across the Bering Strait, with white, sweetly scented flowers which appear later in spring. Older plants can be divided but this sets them back considerably as both species are very slow-growing, so it is best to leave flowering-size clumps alone and plant seed as soon as it is harvested. Shallow water or permanently waterlogged deep, rich soils are essential. If the plants aren't quite happy at the depth you have selected, they will move themselves up or down with the aid of unusual, contractile roots. (Araceae)

*Lysimachia* All species of this genus are water-loving, but most must be planted so that the crown is above permanently saturated soil. They will tolerate flooding for short periods of time, and will grow in shady locations or even full sun as long as they are kept really wet. Most species have yellow flowers; *L. punctata* bears these on upright stems to about 60 cm high, while Moneywort (*L. nummularia*) is a creeper which will grow in water up to 15 cm deep or on quite dry ground. The flowers on the latter species only appear on those parts of the plant actually out of the water, and it is also available in a golden form ('Aurea') which has bright-yellow, coin-shaped leaves when grown with enough sun. There are a number of other species occasionally available, of which I will only mention the reasonably distinct *L. ephemerum* with its white flowers instead of the usual yellow, and *L. thyrsiflora* which will grow in shallow water. All species can be propagated by seed, but selected forms must be divided. (Primulaceae)

*Lythrum* Although there is some slight doubt as to whether *L. salicaria* (discussed under Australian plants) is actually native, there are also two definite non-native members of this genus in Australia. *L. virgatum* 'Rose Queen' produces an abundance of dense foxtails of purplish-pink flowers in summer whether grown in full sun or semi-shade. Although it will tolerate shallow water, it is best grown in wet or waterlogged soils or even well-watered garden beds. A number of other very floriferous varieties of this species are grown overseas, and some of the more unusual colour forms may become available with time — all are propagated from cuttings. Mediterranean Loosestrife (*L. junceum*) is an attractive, sprawling species with showy, pinkish flowers. However, it has already shown considerable potential as a weed, spreading widely from seed in parts of Victoria. (Lythraceae)

*Marchantia* Two species of these liverworts, which usually look like tongues of molten wax flowing in intricate patterns, are grown in this country. Both spread reasonably quickly across wet soils in shade, and during their reproductive phase break out in great rashes of amazing reproductive structures which look as if they belong on a fungus rather than on any green plant. The names I have used are tentative and may also be out-of-date, but it is hard to identify even the native liverworts, let alone exotic species of uncertain origin. My favourite is *M. calcarea*, with its ribbed, mushroom-like fruiting-bodies; the curious, eye-like structures of *M. geminata* are reminiscent of some of those strangest of worms, the planarians. (Marchantiaceae)

*Mentha* All mints enjoy wet or permanently moist positions to varying degrees, and many will grow in waterlogged soils. Full sun results in the greatest concentration of the volatile oils which are such an important part of their fragrance, but nearly all will grow reasonably well

*Marchantia calcarea*

in shadier locations. Only one variety is truly aquatic, the very fragrant and inedible *M. aquatica* which may be happy in water up to 60 cm deep once established, yet is very weedy and invasive. It is usually listed as a species but is probably a hybrid similar to Eau-de-Cologne mint. As many mints grown are of hybrid origin they are best propagated by cuttings or division, though some will come true from seed. (Lamiaceae)

**Menyanthes** Bog Bean (*M. trifoliata*) is so-called for its strong resemblance to young broad bean plants, but is not edible. This is a wide-spreading creeper for shallow water and waterlogged soils with full sun, yet grows slowly enough that it is not invasive. Although the effect of a dense drift at the edge of a pond is very attractive, Bog Bean is mainly grown for the columns of white, fringed flowers in summer, and is propagated by division. (Menyanthaceae)

**Mimulus** Many *Mimulus* species spread so quickly by seed in wet or moist locations that they are best regarded as weeds; indeed, a number have naturalised themselves in wetter countries. The only relatively harmless variety I have grown (I am still trying to wipe out seedlings from *M. guttatus* and *M. cardinalis*) is *M. cupreus* which forms a fast-spreading mat of smallish leaves hidden under a mass of glowing-red, yellow-throated, snapdragon-like flowers in summer. Propagation is by seed or division, and full sun gives the best flowering. (Scrophulariaceae)

**Miscanthus** The only species of these tall and upright grasses available in Australia is *M. sinensis*, a handsome grass for the water's edge which will not grow well if waterlogged. My favourite variety is 'Zebrinus' (Leopard Grass) with its slightly irregular, golden patches along the leaves; these only appear a few weeks after the plant starts out of dormancy in late spring, but remain unchanged for the rest of the growing season. 'Variegatus', on the other hand, is much like any other variegated grass, with creamy-white striping along the blade. More recently developed German varieties of this genus may become available over the next few years but I have not yet seen them. Full sun is best for colour although some shade doesn't affect this, and propagation of named forms is always by division. (Poaceae)

**Mollinia** Only the variegated form of the Purple Moorgrass *M. coerulea* is grown. It is an unexciting tussocky grass of interest only to people who absolutely must have every grass with creamy-white striped blades. This is a frequent host of the poisonous and hallucinogenic fungus causing St Anthony's Fire, but is unlikely to cause problems unless you plan to eat it or grow your own grains. Although it is found on moors in England it will not tolerate waterlogging for more than a few days, and resents drying out or hot sun. Propagation is always by division. (Poaceae)

**Moraea** Most evergreen *Moraea* species have yellow, iris-like flowers and need moist to wet, not waterlogged, soils. Only *M. spathulata* has been grown in Australia for any length of time, and as the other species are not obviously different they are rarely cultivated outside their native South Africa. This is a very attractive and quite distinctive species well worth growing if you have any fondness for bulbous plants. Although slow to establish, mature clumps flower well in sunny locations; propagation is usually by division. (Iridaceae)

**Mosses** are discussed under *Sphagnum* and *Vesicularia* later in this chapter.

**Myosotis** Water Forget-Me-Not (*M. scorpoides*, previously known as *M. palustris*) is very similar to the common, garden species but the flowers are a bit larger and bluer; only the compact form 'Mermaid' has been available in Australia, rarely growing much higher than about 15 cm, but the taller wild form may also occasionally be available. This is a fast-spreading plant for shallow water and waterlogged soils which will tolerate full sun but looks much better with at least a moderate amount of shade in Australian conditions. It is easily propagated by division yet seed will set under some conditions. (Boraginaceae)

**Myriophyllum** There are dozens of native *Myriophyllum* species which include most of the interesting and attractive Milfoils in the world, so it would seem to be a waste of space growing non-native forms. Even so, there are two which must be mentioned, both of which have weed potential even in southern Australia and should not be grown near natural waters. *M. spicatum* has not yet escaped into the wild in this country, but if it were to do so it could become a serious weed — so don't grow it! *M. aquaticum* (previously known as *M. brasiliense*) has already escaped into natural waters but is unlikely to become a severe problem as there are probably only female plants in Australia, so it can only spread from broken-off pieces. This attractive, blue-green Milfoil has finely divided underwater leaves, as do many Australian species; above water they are more feathery in appearance. Both species will grow in deep or shallow water even in quite deep shade, and are easily propagated from cuttings. Several other 'species' have been offered from time to time, but are incorrectly named and uninteresting Milfoils with unknown weed potential and should not be grown. (Haloragaceae)

*Myriophyllum aquaticum*

---

**Narcissus** Strangely enough, two jonquils occur naturally in moist to wet meadows. Like most water-loving bulbs, they must be dried out a bit once the foliage is dying down. Cultivation is much the same as for most other *Narcissus*, though the flowers aren't ordinary in any sense. *N. cyclamineus* has recurved petals which gives the bright-yellow flower the appearance of a bon-bon or some Cyclamens, while *N. viridiflorus* has spidery, deep-green, fragrant flowers. (Amaryllidaceae)

**Nelumbo** Lotuses are among the most spectacular waterplants, with huge, round waxy leaves which shed water like quicksilver and are held well above water. The flowers are immense and have an unusual fragrance which is best described as 'exotic'. The petals drop after the flower has been open several days, exposing the watering-pot-spout seedpods in the centre, frequently used in dried flower arrangements. Most parts of any lotus are edible; the leaves are used to wrap steamed food, the seeds are a delicacy (if you can crack the shell), and the sliced roots are those strange looking rings with neat patterns of perforations you may have seen as an ingredient in some Chinese meals.

I have spent nearly 15 years learning to grow lotus in cooler climates, and have tried seed and tubers of nearly every variety available under many conditions. After all this time I can say that it isn't all that hard to grow some varieties even in Tasmanian conditions, but you absolutely must have everything perfect for the all-important first season. If a new plant grows without interruption in ideal conditions for that first season, it will require no special attention in later years; otherwise, it may not survive the first winter and will certainly require tender care for some years after.

The three basic requirements for a lotus at any stage are a reasonable area of rich soil, a long growing season in reasonably warm conditions, and as much sunlight as possible. The soil mix is much like that used for waterlilies: around 25 per cent well-rotted cow manure, a good helping of blood-and-bone, and for the rest any reasonable-quality potting or garden soil without lime will do. 'Reasonable' warmth varies from one variety to another so it is essential to select one which will enjoy your conditions, but if you live in an area with unpredictable summers you can overcome this by covering the young plant with a plastic cloche just for that first season.

Although lotuses will grow in quite deep water in warmer climates once they are well established, they are best grown no deeper than 15 cm in southern Australia — at this depth both soil and water can warm up quickly. Lotuses are dormant in winter and couldn't care less how cold it gets at that time as long as they don't actually freeze. As the rhizomes are usually around 10–15 cm under the mud and covered by a similar depth of water, it can be seen that few areas of this country are too cold for them.

Named lotus varieties can only be propagated by division, and to give these the longest possible growing season in their first year, the roots are lifted and divided when the plant starts first growth (usually around September or October). The roots form chains of rather sausage-like tubers with growing shoots emerging where these join; a suitable piece for propagation has at least two growing tips. Both tubers and growing tips are quite brittle and must be handled carefully as if all the growing tips are damaged, the tubers will *never* grow again. Ponds with rounded corners are best as sharp corners may become a trap for the fast-growing shoots, killing both them and the plant. To plant them, press the tubers into the soil so that only the growing tips are showing and then weigh them down with smooth stone so they will not float.

Wild forms can also be propagated from seed at temperatures around 25–30°C. As the seed will happily remain dormant for centuries, break the dormancy by filing through the hard shell until the embryo is only just exposed before planting. Sacred Lotus seedlings may not be worth the space and years required to bring them to flowering size in southern Australia because this species has a very wide range and seeds sold are often from plants which cannot adapt to cooler climates. Also, for unknown reasons, many lotus seedlings will not survive their first winter even though they have grown well and reached a large size in their first season.

There are two species of Lotus; the North American *N. pentapetala* (also called Yellow Lotus, *N. lutea*) is only known in the wild form with yellow flowers up to 20 cm across (however, the form grown in Australia is unusually pale, really more a yellowy-cream). The entire plant is slightly smaller in all parts than wild forms of the Sacred Lotus, but will grow and flower at lower temperatures and so is the best choice for cold areas. Plants may flower erratically for the first two or three years, but once they have settled in properly are very free-flowering.

*Nelumbo nucifera* 'Semi-plena'

The Sacred Lotus (*N. nucifera*) has been grown and revered through all of Asia for centuries, and is also native to northern Australia (for simplicity, the native forms are included in this section). The flowers can be immense on full-sized plants (sometimes over 30 cm across), and the wild-type flower forms are every bit as simple and exquisite as the perfect blooms of *N. pentapetala*, but a number of double and very-double forms are also in cultivation. It is hard to say whether these are an acquired taste, or whether no taste at all is required to admire them. I will say no more about them except that the only adjectives I can think of to describe them are blowsy or cabbagey.

Single Sacred Lotuses come in pink or white, sometimes tinged with a creamy-yellow. Native forms offer a reasonable chance of providing cold-tolerant garden varieties if a large selection of clones from cooler areas could be tested (of course, this would take a lot of pond space!). The flower colour in native forms is usually pink though other forms with darker-pink or pure-white flowers have also been recorded. My current favourite is a cold-tolerant form with white flowers tipped with deep-pink veins. The Magnolia lotus (*N. nucifera* 'Alba') is a large-flowered, pure-white form which can be difficult to establish in colder climates. It is not known where this form originated but it is almost certainly not native.

A semi-double form (*N. nucifera* 'Semi-plena') has more petals than the wild forms yet retains their elegance of shape; it is also very cold-tolerant and flowers (for me) in the open in southern Victoria from January to April. Double Sacred Lotus forms include the unattractively named 'Mrs. Perry D. Slocum' in pink, cream and apricot shades. 'Roseum Plenum' is a dark-pink variation, while 'Shiroman' has creamy flowers fading to white over several days. 'Momo Botan' is a miniature red double. Other forms have been offered by one source; I have not seen them myself, but am told they are minor variations, if they are different at all. (Nelumbonaceae)

*Nuphar* Spatterdock (*N. lutea*) is an interesting but less desirable relative of the waterlily with a strong tendency to weediness. It will grow in deeper and colder water than most waterlilies, even in moving water and shaded locations. Although it requires similarly rich soils for reasonable appearance, Spatterdock takes up more space than most waterlilies and the yellow flowers are disappointingly small, globular and not entirely pleasant in fragrance. (Nymphaeaceae)

*Nymphoides* Marshworts are plants of deep or shallow waters which usually develop floating waterlily-like leaves, and the flowers are fringed or tasselled at the edges. The most common non-native species is the Banana Lily (*N. aquatica*), usually offered in the aquarium trade for fish tanks but also a fully hardy outdoor plant. Grown in full sun, the floating leaves develop an interesting, spongy texture below, but the plant is not otherwise very different from our native species. In winter, Banana

Lilies die back to dormant tubers arranged in a shape similar to a miniature bunch of bananas. The flowers of this species are rather plain and white, though there is supposed to be another non-native species in Australia which has orange flowers and which I have not yet seen. Propagation is by division of the many plantlets which form along the runners; as it is so easily propagated *N. aquatica* may even have some potential as a weed. (Menyanthaceae)

**Onixotis** Star-of-the-Marsh (*O. triquetra*) is a South African bulb from areas which flood in winter and spring, but dry out in summer when the bulb goes dormant. The leaves are very slender and erect and the flower-columns carry small but attractive lilac-blue blossoms. Full sun gives best flowering, and propagation can be from seed or by separation of the bulbs in dormant clumps. (Colchicaceae)

**Onoclea** Sensitive fern (*O. sensibilis*) is so-called because at the first touch of frost the leaves turn brown and wither. This is perfectly normal for this cold-tolerant species with strap-fern leaves; it is an unusual but slow-growing creeper for very shallow water and wet soils. Shade is essential, and new plants can be obtained by division once growth has started in spring. (Aspidiaceae)

**Ophiopogon** Mondo grass (*O. japonicus*) is not really an aquatic plant but will grow in shallow water or wet soils. This low, very deep-green, tussocky relative of lilies forms a dense, dark sward quickly if kept wet without being waterlogged. It comes in two forms: the common one grows to around 10 cm high, and another more straggly form to around 30 cm. Shade from the hottest sun is best for richness of colour, though Mondo grass isn't really too fussy as long as it is wet. Another variety, the quite unreal-looking *O. planiscarpus nigrescens*, is a glossy black and looks more like slender tangles of black plastic shavings than a plant. This species needs full sun for strongest colour and will not grow in waterlogged conditions. All species and forms are easily propagated by division. (Liliaceae)

**Orontium** I am going to run out of superlatives very quickly in describing Golden Waterclub (*O. aquaticum*) as it is one of my absolute favourites. It is an Arum-relative for shallow or deep water and will also grow in waterlogged soils (though it never looks as fine as when it is submerged). Deep, rich soil in a sunny position is needed for best development. The lance-shaped leaves are a waxy near-blue. Water doesn't cling to them, it runs off like quicksilver — a beautiful sight in itself. In deeper water the leaves will float, but in the shallows they are held up high. Just before the first leaves appear in middle spring, the flowers start to appear on bone-white stems which may float or be held above the water. These stems are tipped with rich golden-yellow upright flowerstalks like candleflames. The pea-sized seeds which follow should be sown as soon as they fall off, though older plants can also be propagated by division. This species is edible only after elaborate treatment, and I would not vouch for its health-giving properties as many aroids can be poisonous. (Araceae)

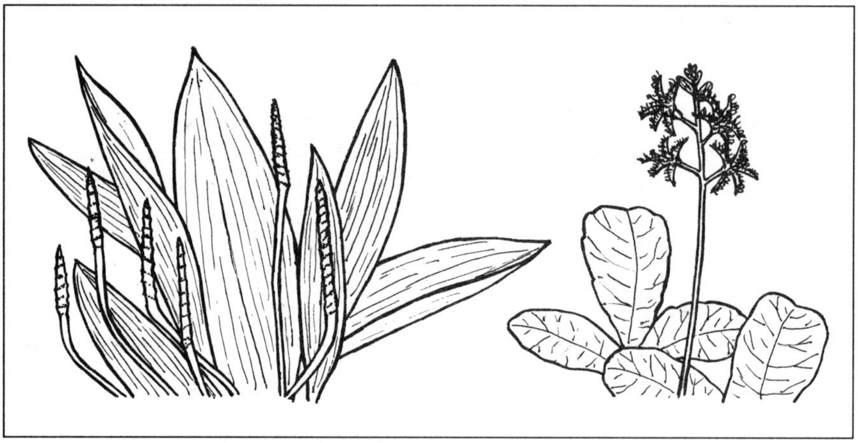

*Orontium aquaticum* (left), *Menyanthes trifolia* (right)

**Oryza** Rice (*O. sativa*) is one of the most important human foodcrops, yet is not easy to grow in cooler climates. Rice needs a long and consistently warm growing season for best production and is particularly sensitive to cold at two stages, when the flower-columns are being formed, and when the flowers are actually opening. In erratic summer climates this gives you two chances to lose the entire crop; however, some short-grain varieties will die back in winter and regrow in warm weather in order to give themselves another chance to set seed. I have had one pathetic plant for three years which repeats this process every year, always dreaming of the day the greenhouse effect will come to the rescue.

A number of rice varieties are much less sensitive to cold at these two stages, particularly some Japanese and Korean cold-tolerant, short-grain varieties, though none of these has as yet been available in Australia. Long-grain and other types are generally much less cold-tolerant, although this varies according to variety.

Rice requires moderately rich soils only. Excessive amounts of nitrogen will cause the plant to become too soft and elongated to support a reasonable crop of seeds. Full sun is essential, and rice is always propagated by seed sown as early in the growing season as possible in colder climates (early September onwards). Transplanting seedlings may set growth back by around two weeks, so direct sowing is best for colder areas. (Poaceae)

*Osmunda* Regal Fern (*O. regalis*) is obviously related to the native King Fern (*Todea*). It is a large-growing waterside fern with deciduous fronds which turn to rich autumnal colours with the onset of cold weather. Regal fern will tolerate some flooding, but does not like prolonged waterlogging; some shade from the hottest sun is necessary in Australia. Propagation can be from spores if you have the patience and necessary skills, or by offset crowns. (Osmundaceae)

*Parochetus* *P. communis* is a dense-growing creeper which loves wet conditions, though the roots will not grow into waterlogged soils. The plant itself looks very similar to clover, but the bright-blue pea-flowers make it quite unmistakable over the warmer months. Propagate by division. (Fabaceae)

*Peltandra* *P. virginica* has upright, arrowhead leaves and greenish, arum-type flowers. It forms dense thickets in shallow water and on waterlogged soils, which could possibly be confused with some *Sagittaria*, but has a distinctive look which cannot be mistaken by anyone with a feel for the delights of the Arum family. Full sunlight or shade are suitable, and plants can be propagated by division or from fresh seed. This species has been eaten in the past, but requires elaborate treatment and is probably not good for the health no matter what is done to it. (Araceae)

*Peltandra virginica* (left), *Peltiphyllum peltatum* (right)

*Peltiphyllum* A curious and very unusual-looking plant, *P. peltatum* creeps across firm soils and clay just above the water's edge and will also grow in any well-watered site. The flowerstems appear before the leaves in early spring, opening out into flattened heads of reddish-pink flowers; the large, disc-shaped leaves with scalloped edges follow soon after. Propagation is by division of the flattened rhizomes in spring; however, plants can be slow to establish and spread, especially if planted in hot or very sunny locations. (Saxifragaceae)

*Persicaria* Rau-Ram or Vietnamese Hot Mint (*P. odoratum*) is very similar to the native *P. decipiens* in appearance, and is a culinary herb used particularly in some Asian cuisines. The peppery, somewhat soapy flavour cannot be confused with any other herb and is definitely an acquired taste. Each long, lance-shaped leaf is marked with a reddish-purple spot. The plant is a sprawling or semi-upright bush which loves wet soil and will tolerate some waterlogging if grown in full sun. Propagation is by division or cuttings. (Polygonaceae)

*Phalaris* *P. arundinacea* is a common pasture-grass and very weedy, but the form 'Variegata' (or Gardener's Gaiters) is ornamental enough to be worth growing in pots to prevent it from spreading. Leaves are variably striped with bright creamy-white (some have almost no green in them), and in spring all new growth is tipped with a rich purplish-pink. All forms of this species will grow in any moist soil or even shallow water. Full sun or quite dense shade makes no difference to variegation, though growth will naturally be slower in shade. Propagation of the variegated form is by division only. (Poaceae)

*Pistia* Water Lettuce (*P. stratiotes*) is a floating, furry-leaved member of the Arum family. It is both attractive and unique, but doesn't survive cold conditions in much of southern Australia. In warmer areas where it grows well, it is a serious weed and usually declared noxious. Water Lettuce will grow in sun or shade, and propagation is uncontrollable if it will grow in your area at all. (Araceae)

*Pontederia* Pickerel Rush (*P. cordata*) is in the same family as Water Hyacinth (*Eichhornia*), and shows distinct similarities in appearance, though without the weediness. The moderately tall, upright and fleshy stems carry glossy, heart-shaped leaves, and in late summer columns of blue flowers rise above the foliage. This is an excellent plant for shallow water or waterlogged soils, as it forms solid and slow-spreading thickets. Propagation is by division, but old clumps may set some fertile seed in warmer years. The seed has been eaten by waterfowl and humans in the past, but doesn't seem to be very freely produced. (Pontederiaceae)

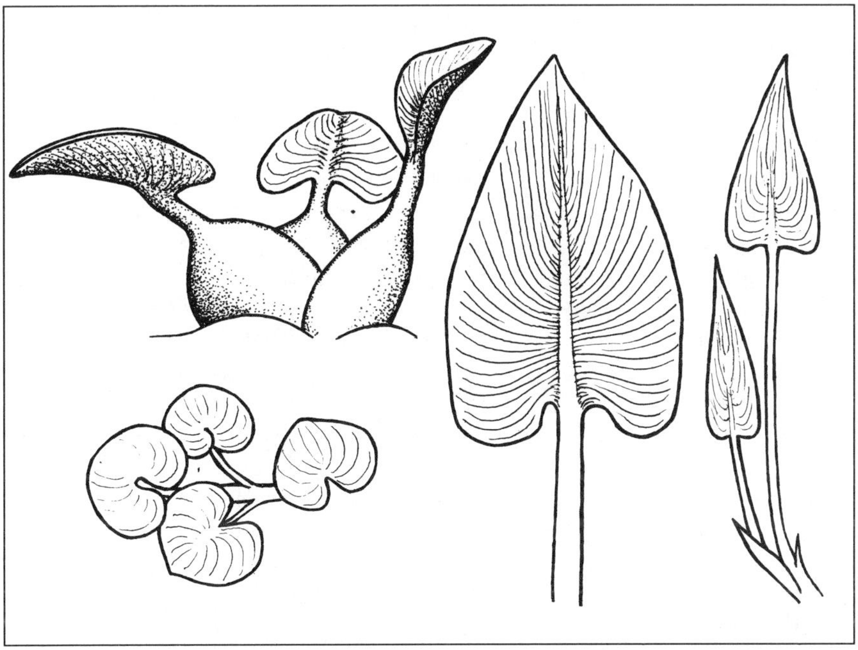

Pontederiaceae foliage shown at approximately a quarter their natural size: *Eichhornia crassipes* (top left), *Heteranthera reniformis* (bottom left), *Pontederia cordata* (centre), and the native *Monochoria cyanea* (right) included here for ease of comparison.

***Preslia*** Strap-leaf mint (*P. cervina*) is, strictly speaking, *Mentha cervina*, as *Preslia* is a synonym for this genus. However, it is listed under the old name in all watergardening books and looks quite unlike a normal mint. The low-growing stems are fleshy and surrounded by peppermint-scented strap-leaves tapering to a point. This species will grow in shallow water or in waterlogged soils and is best in full sun as shade may reduce the fragrance. Divide established clumps for new plants. (Lamiaceae)

***Primula*** Part of this very large genus, the Candelabra and Sikkim primroses, is naturally associated with water. These produce flowers arranged variously in columns above the foliage, sometimes in level tiers which become smaller further up the stem, sometimes in perfect balls. They are all cool-climate plants which must be sheltered from the hottest summer sun, even in Tasmania. All the species can be grown from seed; old seed may germinate irregularly, but in some species, such as *P. rosea*, the seed must be very fresh to germinate at all.

A number of species have been offered in Australia at various times, many of which show every sign of being indeterminate hybrids as this group cross-pollinates freely. The named varieties I have grown have all been correct, however, as they are propagated by division. The descriptions below are of non-hybrid plants as they should appear when they flower. Few species or varieties seem to be as water-loving as they are supposed to be in the cooler climates of Europe and the USA, which I suspect reflects a need for much cooler and better oxygenated water than is usually available in even a mild, southern Australia summer.

A number of species need constant moisture yet start to look seedy if they are wet for too long. These include *P. pulverulenta* whose tiered flowers are red with purplish centres, and *P. viallii* with amazing red poker-heads through which the contrasting pink flowers erupt. *P. denticulata* has spherical flowerheads which are variously coloured in different named varieties, from pure-white to deep-red. *P. japonica* is a strangely-coloured crimson which I have always found quite revolting. (I used to keep quiet about my aberrant response to what is, after all, only a colour but more recently I was delighted to find that Gertrude Jekyll had also found this particular flower-colour unpleasant and knew of many other people who had reacted the same way!)

Other Candelabras may tolerate very wet or waterlogged conditions in cooler climates. These include *P. aurantiaca* (tiered, orange flowers), *P. beesiana* (tiered rose or red flowers with yellow eyes), *P. bulleyana* (deep-orange, tiered flowers) and *P. florindae* (relatively large, pleasantly scented and rather pendulous flowers). The latter species is reputed to grow even in shallow water, though I have never succeeded in keeping it alive under these conditions.

As Primulas are reasonably easy to grow from seed, there are many other species which may become available from time to time. Those suited to moist soils include *P. alpicola*, *P. anisodora*, *P. burmanica*, *P. capitata*, *P. chungensis*, *P. helodoxa*, *P. secundiflora* and *P. waltonii*, while those which will also grow in wet soils include *P. polyneura*, *P. rosea*, *P. sikkimensis* and *P. sinopurpurea*. (Primulaceae)

**Ranunculus** Most Buttercups are waterloving, and many non-native species in Australia are also very weedy, but there are several exceptions.

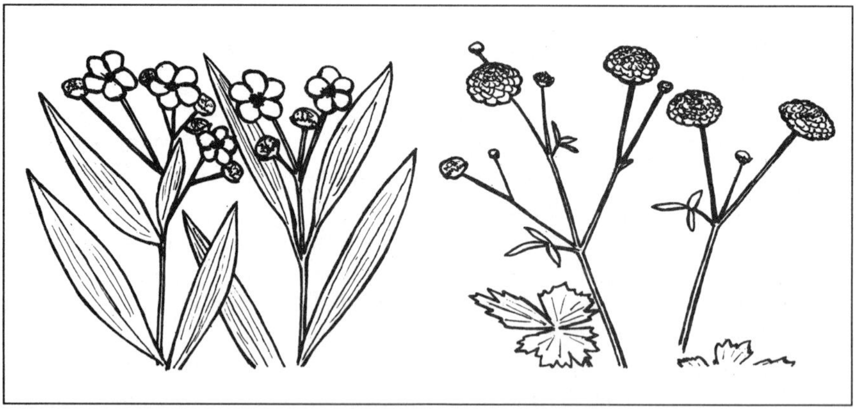

*Ranunculus lingua* 'Grandiflora' (left), *R. acris* 'Flore Plena' (right)

Bachelor's Buttons (*R. acris* 'Flore Plena') is a smaller-growing form with divided leaves suited to very wet or even waterlogged soils, and is covered in masses of smallish pom-pom flowers rather like tiny, yellow, cactus dahlias in spring. *R. lingua* 'Grandiflora' grows in moderately shallow water or on waterlogged soil, with upright, lance-shaped leaves over the warmer months which change to a soft, floppy submerged leaf in winter; the flowers are large but typically buttercup in appearance. This is a fast-spreading species as is the similar but much smaller *R. flammula* which is uninteresting and has enough weed potential to be not worth growing. All species will grow in full sun or with moderate shade and are propagated by division. (Ranunculaceae)

**Rorippa** Watercress (*R. nasturtium-aquaticum*) is excellent as a salad vegetable or lightly cooked and is nutrient- and mineral-rich. *Under no circumstances should it be consumed uncooked from waters containing fresh animal manures* as it may pick up a number of unsavoury bugs, but grown in well-composted or aged manures it is quite safe. Watercress is almost never available from greengrocers because it must have cool, preferably flowing waters for healthy growth and most areas of this country are too warm in summer. If you can't provide such conditions, try growing it in fairly shady locations over the winter months and letting it die away and set seed as the weather becomes too warm. Propagation is usually from seed, which is best planted in autumn in time for the cooler part of the year, but divisions and cuttings also take root easily. (Brassicaceae)

**Rotala** Indian Cress (*R. rotundifolia*) is a creeper for deep or shallow waters. Underwater, the leaves are like those of the closely related *Didiplis*, changing to coin-like chains near the surface. Small bright-pink flowers form small columns above the leaves. This species is tolerant of both low-light conditions or full sun and is easily propagated by division or cuttings. (Lythraceae)

**Sagittaria** Arrowheads are named for the sharp arrowhead shape of the leaves of some species, though others have relatively plain or simple leaves. All species are a bit on the weedy side, but many are so ornamental that I can't bring myself to recommend not growing them — in any case, my recommendations would probably be ignored! *S. latifolia*, *S. montevidensis* and *S. sagittifolia* have edible tubers which must be cooked, the last named species being specially grown as a vegetable in parts of Asia. All species will grow in full sun or in semi-shade, preferably in shallow water though any waterlogged or very wet soil will do.

The species with fairly plain leaves are the least exciting, and include *S. graminea* with a proven track record as a weed — although the leaves can be almost heart-shaped under some conditions they are usually a fairly plain lance shape. *S. lancifolia rubra* has similar foliage, with attractive, red stems which support the leaves, and showy white flowers. Two forms of *S. subulata* have been offered in Australia: variety *subulata* is offered in the aquarium trade as *S. natans*, while variety *kurziana* is a larger, broader leafed and somewhat coarser variation. Both forms have simple, grassy strap-leaves underwater which become shorter and blunter in very shallow water or on waterlogged soils, and are *very* weedy.

Species with arrowhead (sagittate) leaves, on the other hand, are eye-catching foliage plants quite distinct in appearance from most other plants with the exception of some aquatic aroids. *S. montevidensis* is a tall species which has established itself as a weed in disturbed habitats, such as irrigation channels, but does not seem to invade stands of natural vegetation. Even so, it should not be planted near natural waters. The

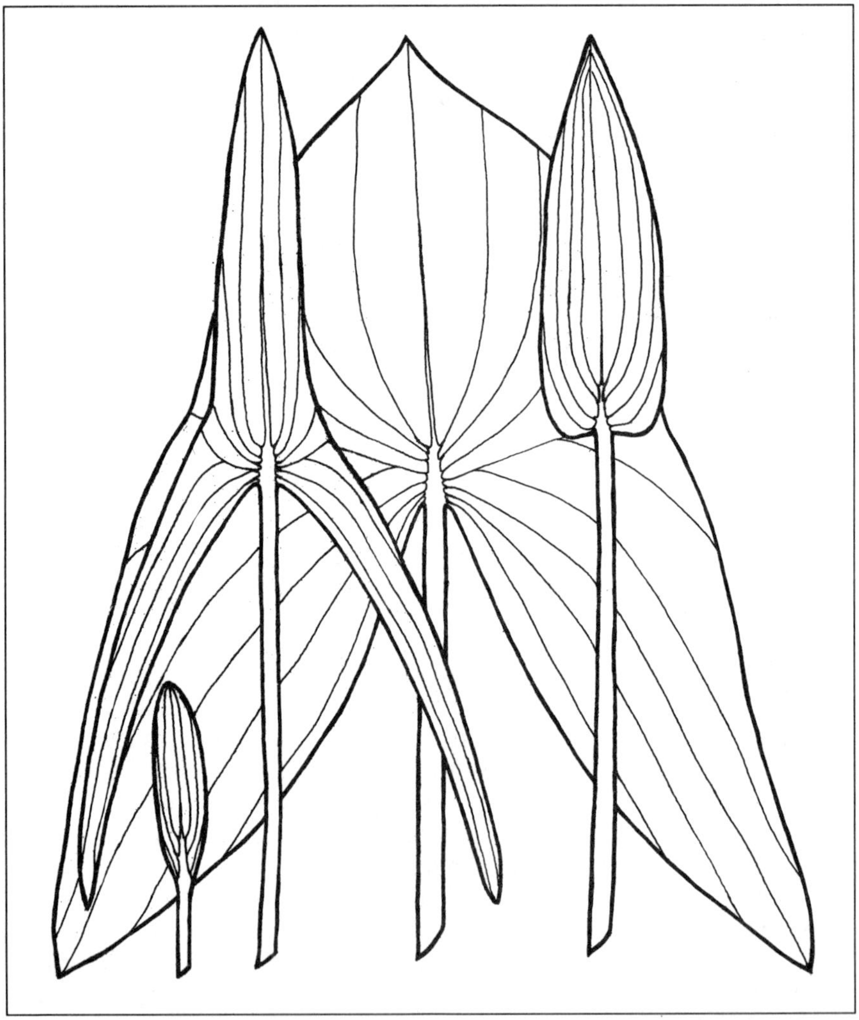

*Sagittaria* leaf shapes: (left to right) *S. subulata subulata, S. latifolia* 'gracilis', *S. montevidensis, S. graminea*

same is probably true of *S. latifolia* which has much more slender and elegant leaves which vary under different growing conditions and from year to year. It may be possible to keep the leaves of this form consistently slender by lifting the dormant tubers each winter and replanting only the smaller ones; flowers of this species are white and quite showy.

Kuwai (*S. sagittifolia*) is grown in many parts of Asia as a more easily cultivated version of the waterchestnut. This is another *Sagittaria* with bold arrowheads for leaves. The large, waterchestnut-like corms are formed without the elaborate treatment required for *Eleocharis dulcis*. Cultivation is easy; just keep growing plants under shallow water for the growing season — the longer the growing season the better the yield. Don't forget that the tubers should not be eaten raw. This ornamental form neither flowers nor sets seed, and is the best choice if you wish to grow arrowheads but don't want to risk having them escape into natural waterways. (Alismataceae)

***Salvia*** Bog Sage (*S. uliginosa*) is an ornamental producing tall columns of sky-blue flowers above the foliage for most of the summer months when grown in a sunny position. Bog Sage prefers moist to wet soils but will not tolerate waterlogging, and is usually propagated by division. (Lamiaceae)

***Salvinia*** All species of this genus of floating ferns are serious weeds outside South America where natural controls keep them within bounds. These are noxious weeds in all parts of Australia and are capable of surviving even light frosts. Although they are attractive enough, in good conditions they will suffocate everything below water by cutting off oxygen with a stifling blanket of weed. The best thing that can be said about them is that they are probably going to be controllable thanks to an expensive biological control programme which has only recently started to have results. (Salviniaceae)

***Samolus*** Water rose (*S. valerandi*) forms small rosettes of rounded foliage in shallow water and grows in shaded locations. I cannot imagine why it is compared to a rose as it is not very inspiring (and is so small that it is quickly buried by larger-growing plants). Propagate by division or from cuttings. (Primulaceae)

***Sarracenia*** These cold-hardy pitcher plants should be cultivated exactly as described under *Dionaea* earlier in this chapter. Each leaf is folded over on itself to form a pitcher which acts as a trap for insects; the trap may be baited with attractive odours or translucent panels which attract light-loving insects. Once inside the upper part of the pitcher, insects are guided by well-placed hairs towards slippery zones whereupon they fall down the pitcher to be digested inside. All *Sarracenia*

*Sarracenia* leaf shapes and positions: (left to right) *S. psittacina, S. minor, S. purpurea, S. flava*

species are efficient predators obtaining all their nutrients from what they trap, and should never be fed any kind of fertiliser.

Lower-growing species include Purple Pitcher (*S. purpurea*) with squat, hairy, purple pitchers. The Parrot Pitcher (*S. psittacina*) seems particularly well adapted to trapping underwater animals when it is submerged, and is therefore very different in appearance to most other species. Other species grow very tall and include the best fly-trapper of all, the Yellow Pitcher (*S. flava*) — I had a single large specimen of this which devoured literally hundreds of blowflies in one growing season until the pitchers were so full that new arrivals could simply walk out of the trap. The best forms of this have golden-yellow pitchers and very unusual flowers of the same colour held well above the leaves. Flowers of *S. leucophylla* are similar although red; this species is mainly grown for its spectacular pitchers which are bone-white with intricate veining in crimson and dark green. There are many other *Sarracenia* species and innumerable hybrids, but the species are usually the most distinctive and interesting in appearance. (Sarraceniaceae)

*Sasa* Only one species of these relatively small bamboos (mostly to around 2 m in height) is very tolerant of wet conditions, the very invasive *S. palmata*. Although it is an attractive large-leafed form, it is also unmanageable in most gardens and will not tolerate waterlogging for any length of time. Some shade is needed for best appearance and new plants are produced by division. (Poaceae)

***Saururus*** Lizardtail (*S. cernuus*) is a low-growing herb with a pleasant odour and heart-shaped leaves. It will grow in full sun or even quite deep shade, in shallow water or permanently wet soils. The flowers are produced in slightly weeping foxtails and are interesting although not exciting. Plants are propagated by division or cuttings. (Saururaceae)

*Saururus*: (left to right) *Saururus cernuus, Houttuynia cordata* 'Court Jester', *H. cordata* 'Flore Plena'

***Schizostylis*** River or Kaffir Lily (*S. coccineus*) in the wild form is a small, Gladiolus-like plant with showy, bright-scarlet flowers. Although normally grown as a garden plant, the natural habitat is the water's edge along riverbanks. Some varieties will even grow and flower well with up to 5 cm of water over the crown. There are many named forms including a particularly good scarlet form called 'Grandiflorus', and a rather small-flowered white has recently become available. However, most named varieties are pink; 'Zeal's Salmon' has large flowers with rather narrow petals, and 'Mrs Heggarty' is a richer pink with broad, well-shaped petals. Full sun is best as shade will reduce flowering, and all named varieties are propagated by division. (Iridaceae)

***Schoenoplectus*** *S.* 'Albescens' and *S.* 'Zebrinus' are discussed under *Scirpus* immediately below.

***Scirpus*** The genus *Scirpus* is now divided into several new genera including *Isolepis*, *Bolboschoenus* and *Schoenoplectus*. All native species are discussed under these names. For the sake of inconsistency, I am listing all non-native species under the old generic name (actually, to keep the nomenclature consistent with the obsolete names used in every European watergarden book). 'Fairylights', now correctly an *Isolepis* species or hybrid, is a fine-leaved, dark-green, weeping tussock which has recently been available as an indoor plant. While it is moderately attractive, only novelty value has kept it around for the year or so it has been available and interest already appears to be fading fast. With many, very similar native species available 'Fairylights' hardly seems to have justified its brief hour of glory, or the trouble of importation. Plants grown in full sun are more compact than those from very shaded positions and their soil should be kept constantly wet. Propagation is easy by division.

The next two varieties are exceptionally choice variegates. I have simplified their names considerably for obvious reasons. 'Albescens' (correctly *Schoenoplectus lacustris tabernaemontani forma albescens*) has cream or green stripes running the length of each stem; these are variable in width so that some stems are green finely striped with cream, while others are cream striped with green. As every combination in between these extremes may occur on the same plant, one would expect the overall effect to be both irregular and tasteless, whereas in fact it is surprisingly subtle and attractive.

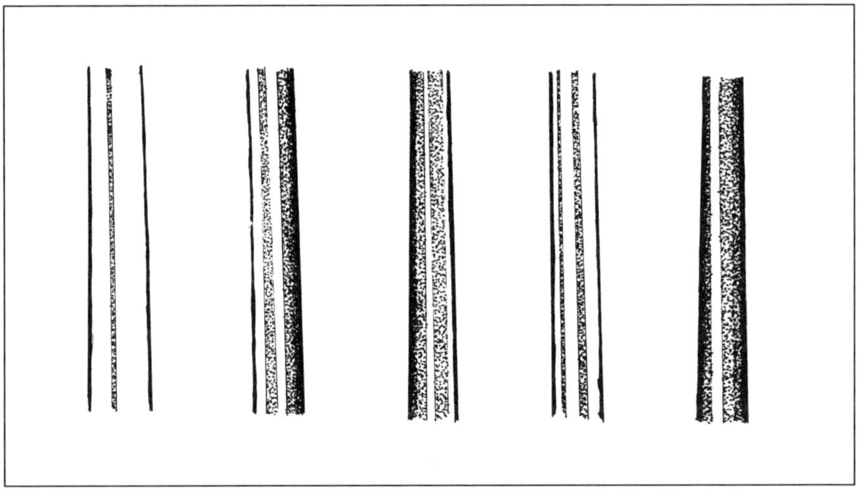

*Scirpus* (correctly *Schoenoplectus*) 'Albescens', showing variations of patterning in the alternating deep-green and creamy-yellow stripes.

'Zebrinus' (*Schoenoplectus lacustris tabernaemontani forma zebrinus*) is the Porcupine-quill rush, a really different variegate with alternating bands of white and dark green along the stems. Some shade from hot sun is necessary if the dark green is not to fade in summer. In this variety, parts of the plant may revert to all-green. Any all-green sections should be cut out to prevent them from overgrowing the less-vigorous variegated sections (such throwbacks must be rare, however, as I have never seen one even though I grow this plant on a commercial scale). Both *Scirpus* 'Albescens' and 'Zebrinus' should be grown in shallow water between 5 and 10 cm deep; propagate by division. (Cyperaceae)

**Shibataea** *S. kumasasa* is a curious, lily-like bamboo which is reputed by some sources to enjoy very wet conditions. Although some degree of wetness does not harm this species, there is a very fine line between comfortably wet and drowning — mature plants can drown in wet sites after several weeks of heavy rain. Although this is a running species, it moves relatively slowly and is effectively a clumping, rather than a spreading, bamboo. Shade from the hottest sun is essential, and division is the only means of propagation. (Poaceae)

**Sisyrinchium** *S. boreale* is a small, iris-like plant with small and rather plain yellow flowers. As it spreads very freely by seed it is best avoided if you don't like weeding or if you do not have the patience to cut off all the seed-pods through the long growing season. Full sun in very shallow water through to wet soils is ideal, and propagation is (of course) by seed. (Iridaceae)

**Sium** Skirret (*S. sisarum*) is related to celery and similar in general appearance. It is grown for its clusters of long, whitish tubers which are fairly sweet in flavour, but must have the stringy central core removed before cooking. Some plants have much less core than others and it is these that should be selected for further propagation by division if you have grown your original crop from seed. Full sun and rich, wet soils produce the best crops; plants will survive waterlogging and even flooding for weeks at a time, though this will slow their growth considerably. (Apiaceae)

**Sparganium** Once regarded as a native species, *S. erectum* is an introduced reedy plant which is showing every sign of becoming a weed. The tough fast-spreading rhizomes can puncture even the thickest pond-liners, but it also seems to spread between river systems with the help of waterfowl. (Sparganiaceae)

**Sphagnum** Although many mosses like wet conditions, few are really aquatic or bog-loving. *Sphagnum* (and closely related species) are the water-loving species most frequently available in Australia. They need sunny but wet and very acid, nutrient-poor conditions for best growth as they cannot compete well with any plants able to survive in richer soils. It is impossible to name any species (even ignoring the difficulties of identification) as imported peat may have added a wide range of alien species from the spores it contains.

**Symplocarpus** Skunk Cabbage (*S. foetidus*) has arum-like, drab foul-smelling flowers which make it a plant only for those who like that sort of thing. Having said that, I must now admit to it being a special favourite of mine. It requires waterlogged or wet soils with reasonable shade from the hottest sun, and is propagated from seed as division will set back even well-established clumps. As a result, this is an expensive species suited only to connoisseurs with unusual tastes. (Araceae)

**Thalia** Water Cannas are tall-growing, dramatic foliage plants closely related to the prayer plants (*Maranta* and *Calathea*). Unlike those genera, they are relatively cold-tolerant plants for sunny sites in water-logged soils, and will grow in quite deep water in warmer areas once the plant has reached a reasonable size. The leaves are large and Canna-like, held on stems up to 2 m high in larger forms, and are followed by even taller columns of purplish flowers. The common form is the blue-grey *T. dealbata* which is also the largest of this genus in Australia. A smaller form of the same species has been offered as *T. dealbata* 'Japonica' but is nowhere near as dramatic and takes up a similar amount of space despite the lower growth. Yet another form of *T. dealbata* has been offered as Red-stemmed Thalia, distinguished only by a slightly purple-blotched stem which may be the result of virus (the true Red-stemmed Thalia is *T. geniculata* 'Rubra' which has not yet been commercially available in Australia). The common form of *T. geniculata* is not as

*Thalia dealbata*

tall-growing as *T. dealbata*, and the lime-green foliage is another obviously different feature. *I. geniculata* 'Rubra' is a selected form from the wild with red stems; the colour doesn't extend into the leaves. (Marantaceae)

***Trapa*** Water Caltrops or Floating Waterchestnut (several *Trapa* species) are not available in Australia, and are a banned import. Illegally imported seed has been brought into this country several times but has not been successfully germinated for reasons I will not disclose as this plant has serious potential as a weed. These species are only mentioned here because the edible nuts have attracted some attention, but they are not a productive food crop. (Trapaceae)

***Trollius*** Globe Flowers are similar to some *Anemone* or *Ranunculus* in appearance, but the flowers are rounded, almost spherical globes. Although they occur naturally in very wet meadows in cold parts of the northern hemisphere, they are not tolerant of warmer waterlogged conditions. In Australia, all forms should be planted with the crown well above waterlogged soils and be kept constantly moist. *T. europaeus* (yellow-flowered) and *T. ledebourii* 'Golden Queen' (orange-yellow) are always listed as species when offered, but are both almost certainly hybrids. *T. yunnanensis* (a paler yellow sometimes with greenish tinting at the base of the petals) is the only true species available. Full sun or light shade are acceptable, and propagation is usually by division. (Ranunculaceae)

***Typha*** By calling native *Typha* species Cumbungi, I avoided the complicated tale which, in order to be thorough, must now be told. *Typha* is now referred to as Bulrush or Cat-tail worldwide, but the name Bulrush originally applied to some *Schoenoplectus* species (which were called *Scirpus* at the time). Books on watergardening in the northern hemisphere have copied this tale from one another for a hundred years to make the point that the real bulrush is actually *Schoenoplectus* (except that they still think it is *Scirpus*). Meanwhile, almost every English-speaking nation has taken to calling *Typha* Bulrush, so let us throw in the towel and surrender to a linguistic deed done a hundred years ago, and call a Bulrush a Bulrush. If you can't cope with English as she is now spoke, just grow the native species and call them Cumbungi.

*T. latifolia* is simply a large-growing rank weed which is no improvement on the equally weedy native, *T. orientalis*. *T. laxmannii* (often still sold as *T. stenophylla*) is a very slender-leaved and relatively slow-spreading species with small, rusty-brown bulrushes; it is a lovely, small-scale bulrush which can easily be kept within bounds. My own favourite is *T. minima*, which grows to no more than 50 cm high with bulrushes so short that they are just little balls on the stems — the perfect bulrush for small pools, and not weedy at all in my experience. All bulrushes are easily grown in full sun in shallow water or wet soils, but they can spread easily by windblown seed and should not be planted near natural waters. Propagation is usually by division, unless you are careless enough to let the seed run loose. (Typhaceae)

*Typhonodorum* A giant version of *Peltandra* which can grow nearly 3 m high, this is the perhaps the most distinctive plant of the Arum family which is truly aquatic. I have only grown it for one winter at the time of writing, yet it seems remarkably hardy for a plant from the tropics — I can remember the newly planted divisions with a 5 cm thick blanket of hail around their stems in the first week I had them. As they have survived the coldest part of that winter and are now growing despite an unusually cold and wet summer at the time of writing, it seems that *Typhonodorum* may be fully adaptable to life in the wilds in southern Australia. Whether they will reach their full potential height cannot be said at this stage, but keep them wet at all times (or even partly submerged in the warmer months) in full sun or light shade. Propagation is usually from the seeds which may germinate while still attached to the parent plant in warmer climates; the seeds are edible but I have no information on whether or not they must be processed. (Araceae)

*Veronica* *V. beccabunga* is a round-leaved creeper for shallow water or wet soils. The small blue flowers are pretty enough, but this species spreads quickly and should not be planted where it can get away. Young shoots of *V. beccabunga* have been used as a salad vegetable. Full sun or shade are equally acceptable, and propagation is usually by division or cuttings. (Scrophulariaceae)

*Vesicularia* *Vesicularia dubyana* is an aquatic moss used mostly as a breeding medium for fishes in tropical aquaria, but seems to grow quite well outdoors in fairly frost-free locations. Unfortunately, it is quickly clogged with algae in the shady, moderately nutrient-rich locations it prefers, and is best kept for low-light conditions in aquariums where few other plants will prosper. (Vesiculariaceae)

*Viola* Most violets love wet or moist positions, but few species are associated with swamps. Of these, only *V. cucullata* has been available in Australia. Like many other plants from cold climates in the northern hemisphere, it seems relatively intolerant of warmth when waterlogged and is best grown as an ordinary garden plant in moist shade. (Violaceae)

*Wachendorfia* *W. thyrsiflora* is an evergreen, bulbous plant from South Africa with some weed potential. From the corrugated, palm-like rosettes of leaves, columns of yellow flowers erupt to 2 m in height, and the seed pods which follow contain copious amounts of fertile seed which will germinate readily in any moist position. Moist to waterlogged soils in full sun are all suitable. Propagation from seed is easy, though mature clumps can be divided if you are very strong and have a large and sharp machete. (Haemodoraceae)

*Zantedeschia* The common white-flowered Arum lily associated with funerals is *Z. aethiopica*, a wetland species which grows and flowers best in wet conditions or even shallow water. Apart from the common form, there is a smaller version called the 'Crowborough' form. 'Green Goddess' is a large plant like the common form, but every flower has a broad border of green very similar to the leaf colour; this is from a selected wild population rather than a cultivar and may come true from seed if not given the opportunity to hybridise with the pure white forms. All varieties will grow in dense shade or even full sun if kept really wet and well-fed. Propagation of selected forms is only reliable by division as naturalised populations are so widespread that self-pollination is uncommon. (Araceae)

*Zizania* There are several Wildrice species and forms, of which only *Z. aquatica* is available in Australia at present. Parched and smoked Wildrice seed is sold at remarkable prices as a gourmet food, but the price is high mostly because this is not a very productive grain and requires much work to collect. Higher-yielding varieties (which don't drop their seed-head at a touch) of similar species are grown in California, and have been selected for better yields and more reliable harvesting; understandably, the growers are not keen to part with live seed to potential competitors. Even so, several of these varieties are already in Australia and should become available in the next few years.

This is an annual species which requires a reasonably cold winter for seed to germinate. The original seed I received from North America included instructions to keep it chilled in a refrigerator until planting time, but I have since had masses of volunteer seed germinate in nursery ponds where it was not originally planted, so it would seem that average winter water temperatures of around 10°C. are probably adequate for self-perpetuating populations; indeed, I am starting to be concerned that the wild form might potentially be a weed! Full sun and depths between 20 cm and 10 cm are ideal, but plants will grow and set seed on drying mud if the water has dropped down to this level towards the end of the season. Seed ripens from the middle of February into April, (depending on when it is planted) and a growing season of four to five months beforehand is optimal for most of southern Australia. This is a tall and striking species sometimes grown as an ornamental in the northern hemisphere, with pink, reddish or lime-green flowerheads. (Poaceae)

# Chapter Five

# Waterlilies (*Nymphaea*)

## Introduction

Waterlilies are the first plant any watergardener thinks of for a new pond, and for good reason; they are easy to grow, many are spectacular by any standards, and a suitable mix of three to five varieties will provide flowers from October right through to June. The colour-range is extraordinary for a single type of cultivated flower, and includes blue, purple, crimson, pink, yellow, orange, cream, and the purest whites, with unusual bi-colour combinations including blue with pink or green, or pastel orange and gold. The flowers vary greatly in size, from as little as 3 cm across, to as much as 45 cm across for some tropicals (grown in the tropics, of course!).

Waterlilies divide naturally into two main groups: the hardy varieties are undisturbed by temperatures far colder than any Australian garden experiences, while the tropical varieties need a long, hot season to show best colour and size of flower. The two groups will not cross-pollinate, so colours cannot be transferred from one group to the other in a hybridisation programme. It is likely that all waterlilies have edible stems when these are peeled, and the rhizomes are also edible once cooked to destroy nupharine, the mildly poisonous chemical they contain.

Hardy varieties produce flowers in a slightly more limited colour range than the tropicals; pure white, cream, yellow, all shades of pink, and crimson reds. Tropical varieties cover much of this colour range, plus the blues, purples and related shades lacking in the hardies (however, the tropicals lack the intense crimson-reds of the hardies). Fortunately for the southern Australian gardener, some excellent tropical blues are quite hardy without winter protection, so every colour can be grown without special treatment in southern Australia.

In a warm year, many hardies will flower continuously from October to April. The tropicals will only start once the water is warm enough for their liking (usually November or December), but will keep flowering until June or even July. This late flowering may be the reason that many varieties will not overwinter successfully in southern latitudes, as their energy is spent on producing flowers until the last moment instead of setting winter-dormant tubers.

## Planting

To grow waterlilies, use any reasonably good-quality soil as your base; for three parts of this, add one part well-rotted cow manure plus about one handful of blood-and-bone for every three or four ordinary buckets (about 30–40 litres) of mix. Planting into a pot is the easiest way to establish a waterlily. Squat pots are usually used because they don't lift the lily too close to the water's surface, but in deeper pools it may actually be better to use a standard pot in order to raise the lily into shallower, warmer water. The plastic mesh baskets used and recommended in older European watergardening books have been available in the past but are a waste of money (they are expensive); I haven't seen any difference in growth between otherwise identical waterlilies in solid and open-sided pots. However, it is possible that the open sides of mesh baskets have some unknown benefit in climates much colder than those of southern Australia.

European books recommend growing hardy waterlilies in full sun, but in southern Australia a moderate amount of shading will not worry most hardy varieties; some will even prefer this to the full impact of our summer sun. Tropical varieties all do best with full sun, but again, many will tolerate some shade without affecting their flowering. Neither hardy nor tropical waterlilies like moving water very much, so avoid using strong fountains or filters in a pond intended for waterlilies.

I find that the easiest method for starting waterlilies in a pond or dam is to allow them to establish properly in a squat pot of 20–45 cm diameter for one season, depending on how large the selected variety will grow. Once the roots have grown right through the soil in the pot, the entire rootball can be emptied out into a dam or lake (assuming that you wish to plant them out) at a suitable depth, and be allowed to spread as the plant sees fit. On steep dam walls, place one or two short stakes below the rootball or the plant may slide into much deeper water where it will quickly lose interest in flowering and life generally. In a pond, leave the plant in the pot until it is time to divide and re-pot. For best flowering, the lily should be lifted and divided every two to three years, less often if allowed to establish in a larger pot before being turned loose in a dam. Most waterlilies left undivided for too long produce masses of densely packed leaves and relatively few flowers, while others may continue flowering fairly well although it is hard to tell under the mass of foliage.

Elaborate classifications have been worked out for the depth at which various waterlilies should be planted, though most will grow and flower perfectly well between 30 and 50 cm below the surface. Some varieties prefer shallower water — from as little as 10 cm for true miniatures — while very vigorous varieties will grow in water up to 2 m deep once they are well-established. In most cases, it is best to start a newly planted lily in relatively shallow water where warmer conditions will produce faster growth, then move it into deeper water — tropical lilies particularly benefit from the extra warmth in the earlier stages of growth.

## Propagation

To propagate a hardy waterlily, divide off single, large pieces with just a single crown or bud at the end and a reasonable growth of roots. Some of the more compact hybrids will produce a knotty ball of rhizomes, roots and crowns which can be difficult to separate out into plantable material. Others produce long rhizomes with easily separated buds and require more space for their vigorous growth.

Tropical waterlilies usually produce rounded rhizomes which look a little like a bulb; larger old ones may not grow or flower very well in southern Australia, but often produce smaller 'bulbs' around their neck. Although some of these smaller 'bulbs' may be smaller than your little fingernail, they can grow into enormous plants with dozens of flowers in their first growing season if grown in large pots. A few tropicals are viviparous (live-bearing), producing new plantlets where the stem and leaf join. In cooler climates they may only produce one or two plantlets, but in the tropics hundreds may be produced.

If you want healthy, named varieties of waterlily, *never* buy pre-packaged tubers; half of these are rotten before they are bought and the others are often so damaged by the treatment they have been through in reaching you that it may take them years to flower. A number of nurseries have also developed the idiotic practice of selling waterlilies by colour alone, on the basis, presumably, that any vaguely similar varieties can be lumped together as single 'colours'. As the colour descriptions are very poor (for example, all the 'reds' I have seen sold without a name are pink), named varieties are a better buy and even the best ones aren't much dearer than the anonymous rubbish.

At this point it is worth mentioning fasciation, a problem which is extremely rare in Australia. This is a dramatic, disease-like condition in which different parts of the leaf grow at different rates, giving a very distorted appearance. There is no cure but to destroy the affected parts; however, buds taken from the apparently diseased tissues revert back to normal growth and so there is no need to destroy the entire plant. This condition is believed to be non-contagious.

# *Hardy waterlilies*

A very large proportion of hardy waterlilies available in Australia were developed in France by Joseph Bory Latour-Marliac around the turn of this century. He is frequently described as a sort of genius, creating in a stroke the most diverse range of colours and textures imaginable. Around the second decade of this century, he offered his secret methods of breeding hardy waterlilies to an English watergardener, Amos Perry, for a sum which only a present-day millionaire could afford. There is some reason to think that his secrecy was intended to add to the resale value of his methods; he was certainly an excellent self-publicist. Many of his varieties were named after famous gardening figures, editors and writers of the time — no wonder they were so freely publicised for him even if they were no real improvement on older varieties. However, Marliac's hybridising techniques were probably relatively simple, and the mystique associated with his breeding programme dates back to a time when no-one else had even imagined the diversity he later achieved. Over years of patient work, he assembled a collection of species and forms which most modern hybridisers of waterlilies have not even seen — these formed the breeding stock from which he developed most of the forms and colours now available.

Even now, there are relatively few really good, hardy waterlilies which Marliac did not originate, or at least develop the parent stock for. However, this doesn't mean that the methods used are a magical secret. We may never know the exact parentage of his early hybrids, but we do know that many are relatively sterile (for example, the Marliacea-types will only set seed sporadically) so further hybridisation using these can waste a considerable amount of effort. Even so, occasional surprises in the form of fertilised seedpods do occur, though if anyone has found out how to achieve this with any regularity they are keeping it a secret. Other hardy waterlilies such as the *N. odorata* species-group (which are often selected forms of wild plants) are very fertile and form the basis of many current hybridisation programmes. For this reason there has been a proliferation of pink and reddish-pink waterlilies in recent years. Let us hope they don't all make their way into this country to clutter up the catalogues of watergarden nurseries.

Many of the older varieties of waterlilies were never clearly illustrated or defined; certainly not in sufficient detail for anyone to be certain just which name belongs with which plant in some rather similar varieties. Further confusion in names has occurred since the original release of other varieties, with names either being misplaced or jumbled; and in at least one case (around 1925), a deliberate substitution being made. The International Waterlily Society is presently embarked on the massive job of sorting out this confusion, defining each variety clearly and unequivocally with a series of botanical paintings plus a checklist of dozens of distinctive characteristics. Once their checklists have been completed it is possible that some of the varieties described below will need to be re-named.

If you would like to try hybridising your own new varieties, there are some hints that I can offer. Allow a quarter of a hectare of shallow, sunlit water (even just 500 or so small ponds will do) for raising the healthier looking seedlings to flowering-size over three to five years. Nearly all of these will be no improvement on the varieties already available as most will either show little vigour, flower poorly, or will simply be boring little weedy things. Good results take time and patience, and though a number of varieties have been bred in Australia, few will stand the test of time.

Only two varieties of hardy waterlily which have been raised in Australia are well-regarded overseas at the present time. 'Norma Gedye' is an unexciting pink which was probably an accidental seedling rather than the result of a deliberate breeding programme. 'Moorei' is an old variety selected at Adelaide Botanic Gardens, basically a slightly richer yellow version of 'Chromatella'; recently this has become the single most popular waterlily available in Europe, though no-one quite knows why. One Australian watergarden nursery has recently claimed to have raised a large number of distinctive, hardy waterlilies (even hinting at improbable crosses using tropical varieties); however, the examples I have seen have not been exceptional. If you are considering one for your pond, I would recommend comparing them closely to older, better known varieties, as prices for some of these 'original creations' have been very high for what are rather ordinary waterlilies.

The varieties described below are mainly those I believe to be the best in Australia, although not all of them are commercially available at the time of writing. Those included have mostly been selected for fine colour, fragrance, reliable flowering, or other special qualities. Also included are a few which, though not necessarily among the best, are different enough that they should be described for the sake of completeness. Fragrance is only mentioned if it is particularly strong or unusual as most waterlilies are at least slightly fragrant. I have used the internationally accepted names wherever possible; these are not always the names these varieties are known under in Australia, therefore the invalid names are also included to avoid confusion.

## White and cream varieties

Many hardy species come in pure white forms, and so the hybrids and selected forms described are a varied lot. A disproportionate number of fragrant forms are included, as many white lilies have almost no fragrance; less-fragrant forms have only been included if they are sufficiently distinctive. All white waterlilies have a contrasting yellow centre.

**'Carolineana nivea'** has a reasonably large flower which may be held above water and is fragrant; this variety suits smaller ponds which are not too shallow.

**'Gladstoniana'** produces really huge, very broad-petalled flowers, but will only grow well in large pools and deeper water.

**'Gonnere'** ('Snowball') is double-petalled and very rounded in shape, similar to a compact dahlia flower. The plant itself is very compact-growing and so is well-suited to small pools.

**'Loose'** has slender, rather curved, pure-white petals on a plant which may hold the flowers above water. This is a very distinctive elegant and deliciously fragrant variety, outstanding among the whites.

**'Hal Miller'** always gives me trouble deciding whether to call this a white or a yellow, which is not surprising because it is actually creamy-white perhaps with the faintest hint of yellow. It was bred from 'Sunrise' to which it is very similar apart from the colour. The big, starry flowers appear on a very large plant and are sometimes held slightly above water. It is another variety best suited to larger pools and deeper water.

**'Marliacea albida'** is the lily most likely to be offered as an un-named white. It is large flowered with a lemony but delicate fragrance, and is a fast grower best suited to larger, deeper pools.

**'Odorata alba'** grows compactly and is excellent for smaller pools, producing medium-sized, very fragrant flowers freely.

**'Queen's Bridge'** is a starry-flowered variety with a scent like jasmine tea. Although it looks most spectacular with room to spread in deeper water, well-established plants will also flower well in surprisingly shallow water. This is probably a selected form of a true wild waterlily.

## Yellow varieties

The yellow hardies derive their colour from *Nymphaea mexicana*, which is not a well-recommended species in its own right. In southern Australia, some of these hybrids may keep many of their leaves over winter (unlike most other colour forms) and will often be the earliest and latest flowering of the hardy varieties. None of the yellows which have been sold in Australia to date are particularly fragrant.

**'Charlene Strawn'** holds large, starry flowers well above water. This variety is reputed to be fragrant but has not yet been released commercially.

**'Chromatella'** has straw-yellow flowers with a golden centre; the leaves are neatly and attractively splashed with purple. This is a slow-spreading plant suited to smaller pools.

**'Moorei'** is very similar to 'Chromatella' but is a slightly deeper and richer yellow. The purple splashes on the leaves are smaller in this variety, more like conspicuous freckles than the purple splashes of 'Chromatella'.

***Nymphaea mexicana*** is a pest which will quickly form choking mats of leaves in warmer climates. In cooler climates it isn't as invasive, but then it isn't very free-flowering either.

**'Sulphurea'** produces smaller and much paler versions of 'Sunrise' flowers (it has been confused with this variety in Australia) and will grow in smaller pools.

**'Sunrise'** makes starry, rich-yellow flowers over 20 cm across in good conditions. The plant is very large and wide-spreading and will also grow in quite deep water once established. The first few flowers in the growing season may look rather rumpled, but later flowers smooth out as the weather becomes warmer.

## Paler pink varieties

The dividing line between the paler pinks and deeper colours is not easily defined, especially as the flowers of many pink waterlilies will turn darker or paler over several days. However, I think it is worth making a distinction because there is a considerable difference in overall effect between the varieties described immediately below, and those with much deeper colours.

**'Amabilis'** is a very large-flowered variety with sharp-looking petals, shifting from salmon-pink to a rosy colour.

**'Arc-en-ciel'** flowers are a soft pink, though this variety is grown mainly for the foliage as it flowers rather poorly. The leaves are splashed with various shades of purple, pink, bronze and white which fade as the leaves grow older.

**'Carolineana perfecta'** is actually 'Odorata W. B. Shaw', a mistake which goes back a long time in Australia. The true variety is also in Australia which may cause some confusion when it is released unless some growers can be made to see the advantages of using the correct names for waterlilies.

**'Gloire du Temple sur Lot'** has unusual, very-double flowers which open paler-pink, then fade to almost-white with age. The fragrant blooms may have as many as a hundred petals and have been compared to chrysanthemums floating on the surface of the water. Marliac must have been particularly proud of this variety as he named it after his own home; however it may take years to establish and is very shy-flowering. Personally, if I wanted chrysanthemums that is what I would grow — they flower much more freely!

'Gloire du Temple'

**'Marliacea carnea'** (Morning Glory) is a freely flowering and very fragrant variety which looks almost white from a distance. However, from close up it can be seen that each petal is flushed with a delicate yet rich shell pink.

**'Odorata W. B. Shaw'** produces good-sized, beautifully shaped, open flowers of a pale salmon or shell pink with a golden centre. Not only is this variety very fragrant, but the scent is also one of the most delicious I can imagine. A number of other writers on watergardens regard this as the finest of waterlilies (and if I had to make a choice I might agree). This form self-seeds freely, hence some inferior, almost-white versions have also been offered under this name.

## Deeper pink varieties

Various shades of pink are perhaps the most common colours in hardies, and so the list which follows includes an assortment of very different waterlilies. Some of these are sometimes described as red for no apparent reason. If you are in any doubt as to which deep pink variety most appeals, choose 'Rose Arey' — I don't know of anyone who has been disappointed by it.

**'Arethusa'** is a large-flowered variety, deep-pink shading towards crimson in the centre. The outer petals are reddish-tipped.

**'Firecrest'** makes fragrant, medium-sized flowers, open in shape with the petals rounded at the tips; the centre is orange. This variety is best grown in small pools, and will stay quite compact.

**'Hollandia'** is free-flowering with large, fully-double flowers, opening medium pink and darkening with age.

**'James Brydon'** is a prolifically flowering variety covered with dark-pink, cup-shaped flowers and an apple-like fragrance. The plant itself is a compact grower well-suited to smaller ponds.

**'Pearl of the Pool'** is slightly double with good-sized, medium-pink fragrant flowers. It is free-flowering.

**'Perry's Pink'** is similar to 'Hollandia' but deeper in colour. I have only seen this variety on slides but it is supposed to be exceptional among pink doubles.

**'Pink Sensation'** has large flowers resembling a paler, more rounded 'Rose Arey', and with a similar fragrance. This variety is supposed to open early and close late, but the difference isn't great under Australian conditions.

**'Ray Davies'** was chosen as the best hardy variety by the International Waterlily Society in 1987. This is a large, double-petalled flower, not all that different from other good doubles, which gives some idea of the difficulty waterlily breeders have had in producing something really new and striking.

**'Rose Arey'** flowers freely with starry, rich-pink flowers, and is a fairly moderate grower which will do well even in smaller pools. The fragrance is both strong and special, something like a blend of aniseed and honey.

A number of watergarden experts regard this as the best of hardy waterlilies, and while I don't agree, they certainly aren't far wrong.

**'Rose Nymph'** is a free-flowering rose variety which is fragrant.

**'Somptuosa'** produces distinctive deep-pink flowers mottled with white, which unfortunately gives them a slightly blotchy appearance sometimes.

## Red varieties

All red varieties have white or partly-white outer petals (although strictly speaking, this outermost ring is made up of sepals). In the more richly coloured varieties this gives a pleasing contrast which emphasises the depth of the red colouring towards the centre of the flower. In less intensely red varieties the effect is sometimes rather patchy looking, especially as many of the red petals in such varieties are white-tipped. Newly planted or divided plants sometimes produce several pink flowers before these change to the full colour — generally closer to crimson than scarlet. (Pink waterlilies may also show considerable variation in colour early in the growing season, but not to the same degree.)

**'Atropurpurea'** has medium-large, open flowers with a golden-orange centre, and is reasonably fragrant.

**'Escarboucle'** is very large-flowered and a fairly uniform deep-red which is best suited to larger pools because of its exuberant growth, flowering even in quite deep water once well-established. Some water-gardeners regard this as the best of the hardy waterlilies; I strongly disagree as the petals are often rather rumpled in appearance, and the flower is so open that it can look as though it is falling apart. This variety looks and grows best in large expanses of water, and is not suited to the smaller pools found in home gardens; however planted in a dam and seen from a distance it is striking.

**'Froebeli'** is a medium-sized and rich-red selection of *Nymphaea alba* var. *rubra*, highly regarded in Europe though probably not tolerant of our high summer temperatures.

**'Gloriosa'** has medium-sized flowers which are reasonably uniform in colour, yet are rather pinkish in the early part of the flowering season and never reach the same depth of red as most other varieties. The plant is small-growing and will do well even in small or shallow pools with some shade.

**'Laydekeri fulgens'** has deep-red flowers with a contrasting white surround, and flowers freely. This is another variety which may not appreciate the heat of Australian summers.

**'Newton'** holds large, starry flowers well above water. They are quite deeply coloured, with whitish tips, but the overall effect is not exceptional.

**'Vesuve'** is the darkest purplish-red waterlily I have seen, partly because the centre is rather dark unlike the brighter colours of most other red waterlilies. It should be grown in deeper, cooler water and will not do well in areas which are hot for most of the summer.

**'William Falconer'** has large, fragrant flowers which can be very dark in colour. This is my favourite of all the red waterlilies, perhaps even of all waterlilies. The petals may be completely red and are beautifully offset by a ring of near-white sepals; the centre of the flower is a luminous combination of glowing orange and golden-yellow. Unlike many red varieties, this one will flower well even in quite hot climates.

'Vesuve'

## Changeable or sunset varieties

**'Sunset' hardies** are often described as 'interesting', which is a polite way of saying they aren't very colourful. They are variations on yellow and apricot, usually changing over several days to deeper, more coppery colours which may include orange or red. The colours vary considerably in different climates, so it is possible that we in southern Australia never see them at their best.

**'Comanche'** opens yellowish but changes to apricot, and is otherwise unremarkable. 'Comanche orange' is another variant and is much more apricot in colour, though with a reddish tint. It is best used as a variation in large groupings of differently coloured waterlilies, and looks most attractive from a distance.

**'Phoebus'** is quite cup-shaped, opening a pale icy yellow and darkening over several days. The centre of the flower is reddish-tipped. This may not be the true 'Phoebus' but it seems quite close to the original description.

**'Sioux'** has slightly starry flowers opening yellow-apricot and becoming more coppery with age.

## Miniature varieties

The varieties listed below include all those varieties sold as miniatures in Australia, though some are much larger than the bona fide miniatures. True miniature varieties will grow and flower best in water 10–20 cm deep; however, the larger plants under this heading will also do well in small ponds and quite shallow water.

*Nymphaea tetragona*

**'Aurora'** is a changeable miniature which I find drab; the colour shifts from an apricot colour to a battered brownish-red over several days.

**'Indiana'**, as defined in Australia, has flowers which are an off-yellow, deepening and developing a faint reddish tint over several days. The plant sold under this name in Australia may actually be 'Graziella' which has much smaller flowers than the true 'Indiana'.

**'Laydekeri lilacea'** has also been offered as 'Laydeckeri Pink' in Australia. This is another variety much larger than true 'Pygmaea' types, with medium-pink flowers flecked with fine, white speckles.

**'Maurice Laydecker'** resembles 'Pygmaea Rubra', but has white-flecked flowers. It is slow-growing (and shy-flowering) and for this reason will probably not be available for some time.

***Nymphaea tetragona alba*** is a miniature species from Asia with pure-white, golden-centred flowers usually around 2 cm across on younger plants, becoming a little larger on more mature plants. As it is a species it is most easily propagated from seed, but division of fairly young plants with distinct crowns is also successful.

**'Pygmaea Helvola'** resembles a tiny 'Chromatella'. The pale yellow flowers (up to 3 cm across) have a golden centre, and the purple-splashed leaves are not much larger. This little gem is free-flowering and can even be grown in a large goldfish bowl in a sunny position. (The fish, however, will probably not enjoy the heat as much as the lily will.)

**'Pygmaea Rubra'** has tiny, red flowers with orange centres; as with other red varieties, the outer petals are rather pale. The leaves are purplish with a red tint. This is a very slow-propagating variety, so it is likely to be expensive when it becomes available and will probably remain so.

# *Tropical waterlilies*

Tropical waterlilies come in a wide range of colours which are distinct from those of the hardy waterlilies. They hybridise easily, and many of the hybrids are fertile. As a result there is potentially a larger range of varieties available than there is in the hardies. However, many nurseries offer fewer varieties of tropicals because of the difficulties in raising them. In northern Australia, tropical waterlilies grow so large and so well that there is no reason for anyone in those areas to plant any of the hardies (which will also grow perfectly well) except, perhaps, the crimson varieties. In southern Australia they are never as fine or as large, and the colours may be paler as well in many varieties.

In general, tropical waterlilies produce larger flowers than hardy varieties if they are grown in a sufficiently large container, but the maximum size of the flower usually given in catalogues refers to plants grown in the tropics — the cooler the climate, the smaller the flowers will be (although cold-tolerant varieties don't seem to vary so much). Flowers are usually held well above water, and all varieties are fragrant, with scents quite distinct from those of the hardies.

Soil mixes and general cultivation are no different to the hardies, except that you should use larger pots and allow more space. Tropicals are usually planted as tubers often as small as a fingernail when they start into growth between October and December. Despite their small size, they will grow into enormous plants within several months, particularly if put into a pot of 30–50 cm diameter. Once they have started flowering, they will not stop until their foliage is killed by cold weather in late autumn or early winter. Unfortunately, this dedication to flowering may mean that they don't put enough energy into growing new tubers. Perhaps for this reason, many tropical lilies won't reliably survive winter outdoors in southern Australia, or will become woody and choke themselves out within a couple of years

The tubers can be lifted when foliage dies down and are best stored in moist sand or peat in a cool place until it is time for replanting in late spring or early summer. Relatively cold-tolerant varieties can be left to overwinter in 30–50 cm of water. Be sure to move their pots into the shallows in spring to encourage new growth. However, most tropical waterlilies tend to develop very large tubers which don't grow or flower well in their second year, and it is best to plant small young tubers of these varieties each year for maximum flowering.

You can produce your own young tubers, though this requires some organisation. Older tubers may produce new ones around their 'neck', or you can pinch-off plantlets from the main tuber in spring and grow them on. Live-bearing varieties develop new plantlets at the base of older leaves in warm climates, more rarely in cooler climates. In southern Australia it may be two years before a flowering-size tuber is produced, so several plantlets should be maintained each year, choosing the best one when you are ready to grow it up to flowering-size.

The selection of varieties that follows includes several which don't require the elaborate propagation described above. However, always go against your instinct at the beginning of the growing season and replant the *smallest* healthy tuber you can find on these varieties; even a tiny tuber may grow into a plant the size of a bathtub in several months and will flower more freely than a woody old tuber.

There are perhaps as many as 100 tropical waterlily varieties which have been sold at one time or another in Australia. On average, they are more uniform in quality than the hardy varieties, though many are still rather mediocre. On the other hand, the differences between varieties can be so minor as to be not worth mentioning. The list below is only a brief guide to the varieties available; mostly I have emphasised those which are likely to survive outdoors in cooler areas of southern Australia. As the colour range of many tropicals overlaps with that of the hardies, the list also focuses on colours and combinations not found in the hardies, particularly the blues and purples. You should also bear in mind that red in tropical lilies is more of a scarlet-orange than the very different crimson-red of the hardy varieties.

## Blue and purple varieties

These are very common colours in tropical waterlilies. The varieties of these colours described below are reliable growers and fairly cold-tolerant and are the best I have seen to date for cooler climates.

**'Blue Beauty'** produces large, medium-blue flowers spotted with black on the outside of the flower. The plant offered under this name in

Australia is quite cold-tolerant although slow to reproduce; however, it may not be the true variety (which is perhaps a much deeper colour).

***Nymphaea 'capensis'*** comes in many attractive forms, none of which are available in Australia. The versions of this species we do have at the present time are fast-spreading weeds in warmer climates, but grow and flower poorly in southern Australia. All of the colour forms I have seen here are quite insipid in colour and pale blue at best, sometimes with a lilac tint. A recent revision of names of tropical waterlilies regards this species as *N. nouchali* var. *zanzibarensis*, but it is too early yet to comment on whether this will be accepted.

***Nymphaea colorata*** is generally called the 'Blue Pygmy' as it is the smallest-flowered tropical in cultivation. The lilac-blue (sometimes darker) flowers reach 10 cm across, very much larger than the true miniature hardy varieties. Although attractive, it does not seem to be particularly well-suited to cooler climates, and *N. stellata* may grow and flower better in similar spaces.

***Nymphaea stellata*** holds its starry, medium-sized, sky-blue flowers well above water. It is a distinctive and fragrant species which is easy to grow and multiply in colder climates; indeed, it could be regarded as completely cold-hardy in most of southern Australia.

**'Royal Purple'** has very large open flowers in various shades of purplish-blue. This is a live-bearing variety with considerable tolerance to cold, especially when kept in deeper water over winter.

**'William Stone'** is a less starry-looking waterlily than *N. stellata* and doesn't have much personality either. It is probably descended from that species and has inherited reasonable cold-tolerance from it.

## Other varieties

The assortment below is not a complete listing of varieties, rather it is a selection of unusually-coloured forms which are reasonably distinct from hardy waterlilies.

**'Afterglow'** is subtly multi-coloured, shading from apricot-orange to pinkish and yellow. The flower may be less sharply coloured in cooler climates where colours fade to a tone not unlike that of an old autumn leaf.

**'Green Smoke'** changes from an aqua-green (with the usual golden centre) to medium blue at the tips, and is impossible to mistake for any other waterlily available. The flower is large and the plant is reasonably cold-tolerant but very difficult to keep alive through winter; it is also slow to reproduce.

**'Jack Wood'** is a little like a raspberry in colour (perhaps with a dash of scarlet) but to my eye seems not much better than the sunset-coloured hardies, at least in southern Australian climates.

**'Persian Lilac'** has deep lilac-pink flowers, a colour quite unusual and different from most of the tropical waterlilies, but not so unusual or interesting when compared to the best of the starry, pink hardy waterlilies.

**'Red Beauty'** is red inclined towards scarlet, but is not exceptionally cold-tolerant.

**'St Louis Gold'** gives a yellow-gold effect which is not all that distinctive and certainly not more interesting than the hardy 'Sunrise'. However, it is relatively easy to grow for a tropical variety.

## Night-flowering waterlilies

Nocturnal-opening tropicals are night-flowering plants opening towards sunset and closing around mid-morning. Unless their short daytime appearances fit well with your lifestyle, you will need good night-lighting for your pond or dam — a hand-held torch is really not the same! They are also more cold-sensitive than many other tropicals, and are mostly grown when good-quality night-lighting is already planned in relatively warm climates. Bear in mind that even in summer, night temperatures in southern Australia can be much lower than day temperatures and this will affect flowering.

There are perhaps eight or ten night-flowering waterlilies in Australia at present, of which the following are good examples: **'Emily Grant Hutchings'** is a glowing coral-pink; **'Red Flare'** has rich-red flowers on red stems; **'Wood's White Knight'** is an interestingly shaped and attractive creamy-white.

All waterlilies belong in the family Nymphaceae.

# *Suppliers of Waterplants*

The list below is not necessarily complete, but gives a good selection of nurseries which offer waterplants to the public.

## Direct sales of waterplants

Amaroo Water Gardens, Bevan Road, Hope Forest, SA 5172

Arcadia Lily Ponds, 151 Arcadia Road, Arcadia, NSW 2159

Austral Watergardens, 1295 Pacific Highway, Cowan, NSW 2081

Gedye's Watergardens, 37–41 Elizabeth Street, Doncaster East, VIC 3109

Ledora Farm, Pacific Highway, Mt Kuringai, NSW 2080

Lillyput Watergardens, Jimboomba, QLD

Lotus Blossom Watergardens, 597 Jarvis Road, Baldivis, WA 6171

Lotus Watergardens, Springvale Road, Keysborough, VIC 3173

## Waterplants by mail order

You can buy waterplants by mail order no matter where you live; they are the easiest plants to send in this way as they are already adapted to long periods without oxygen or light.

Austral Watergardens, Pacific Highway, Cowan, NSW 2081. Probably the largest range of waterlilies in Australia; many other waterplants are also available. Send three stamps for their colour catalogue.

Dragonfly Aquatics, RMB AB 366, Colac, VIC 3250. The largest range of waterplants and wetland plants in Australia, specialising in natives, edible species, fragrant waterlilies, rarities, water-irises and grassy plants. Send two stamps for the catalogue.

Lotus Blossom Watergardens, Stakehill Road, Baldivis, WA 6171. Relatively few waterplants are available in WA; this nursery seems to have the largest range in that state. Send three stamps for a catalogue.

The Watergarden Nursery, PO Box 68, Emerald, VIC 3782. A fair range of waterlilies including several not yet available elsewhere; some other plants are also available, particularly Kaempferi irises. Send $1.50 for a catalogue.

# *Suggested Reading*

There are many books dealing with the subjects touched on in this volume. The books listed below are those which are most widely available (including libraries), up-to-date, and which give good information on their subject areas.

## Ponds

Swindells, Philip, and Mason, David, *The Complete Book of the Watergarden*, Ward-Lock, 1989. This is probably the most comprehensive guide to all the fiddly technical details involved in setting up a pond. However, another book, *The Stapely Book of Watergardens* (authors not named), is almost as good on such details and is more likely to be available in libraries.

## Dams

Nelson, K.D., *Design and Construction of Small Earth Dams*, Inkata Press, 1985. This is the only reasonable and comprehensive book on dam construction in Australia; with 'small' referring to any dam smaller than would be needed to supply water to a whole city!

## Fishes

Leggett, Ray, and Merrick, John R., *Australian Native Fishes for Aquariums*, J.R. Merrick Publications, 1987. This book deals in detail with native fish in captivity, though fish kept outdoors are not covered quite so thoroughly. If you have a strong interest in native Australian fishes, I recommend that you join the Australian and New Guinea Fishes Association (ANGFA). Benefits include a high-quality journal with colour plates, articles by Australia's best-known ichthyologists, and a quarterly Bulletin which keeps you up-to-date on more recent developments. Despite all this, membership is surprisingly inexpensive. Current costs and details can be obtained by writing to ANGFA, PO Box 502, Ringwood, VIC 3134.

Hervey & Hems, J., *The Goldfish*, The Batchworth Press, 1948 (more recent paperback editions are also available).

Matsui, Dr. Y., *Goldfish Guide*, Pet Library, 1972, (recent reprintings are by Tropical Fish Hobbyist (TFH) Publications).

## Water in the Garden

Thompson, Paul, *Water in the Garden*, Lothian Books, 1991. This is by far the best book on this broad subject that I have seen, and would make an excellent (and inexpensive) companion volume for this book, focussing as it does on design.

## Waterplants

Sainty, G., and Jacobs, S., *Waterplants for New South Wales*, Water Resources Commission of NSW, 1981. This book has good coverage of most of the waterplants of NSW and is illustrated with hundreds of colour plates.

Swindells, Philip, *Waterlilies*, Croom Helm, 1983 (paperback editions have been available more recently). A very good coverage of waterlily species and varieties, although many newer varieties are not listed.

## Freshwater life

Williams, W.D., *Life in Inland Waters*, Blackwell Scientific Publications, 1983. A very good introduction and guide to Australian freshwaters (and saltwater life too).

Williams, W.D., *Australian Freshwater Life*, Globe Press, 1980. The most comprehensive guide to the animals living in Australian freshwaters.

# Index